More Praise for *The Seashell*

"Alan Cutler's excellent biography of the [...]
to Galileo is an absorbing study of a figu[...]
gotten."   —*The Dallas Morning News*

"Alan Cutler's fascinating history ably explores how the prevailing biblical explanations were turned upside down by Nicolaus Steno, one of the most undeservedly unknown geniuses in the history of science."   —*The Portland Oregonian*

"Cutler's smart and readable biography puts Steno right at the forefront of the geological revolution."   —*Natural History*

"A sophisticated portrait of a forgotten pioneer."   —*Booklist*

"Like Dava Sobel's *Longitude,* Cutler's highly readable work compellingly depicts the significant discoveries of a single individual who changed prevailing perceptions."   —*Library Journal*

"An exemplary book of its kind: brief, informative, well-researched."
                                            —*The Washington Times*

"[A] strong portrait of an unsung innovator, an intellectual meteor that struck the world of geology and sent it slowly spinning."
                                            —*Kirkus Reviews*

Alan Cutler has a Ph.D. in geology and is a writer affiliated with the Smithsonian Institution. Dr. Cutler was a contributing editor to the book *Forces of Change: A New View of Nature,* a joint publication of the Smithsonian and the National Geographic Society. Dr. Cutler's writing has also appeared in *The Washington Post* and *The Sciences,* among other publications. He lives in Gaithersburg, Maryland.

Visit Alan Cutler at www.alan-cutler.com

The Seashell on the Mountaintop

How Nicolaus Steno Solved
an Ancient Mystery and Created
a Science of the Earth

ALAN CUTLER

A PLUME BOOK

PLUME
Published by the Penguin Group
Penguin Group (USA) Inc., 375 Hudson Street, New York, New York 10014, U.S.A.
Penguin Books Ltd, 80 Strand, London WC2R 0RL, England
Penguin Books Australia Ltd, 250 Camberwell Road, Camberwell, Victoria 3124, Australia
Penguin Books Canada Ltd, 10 Alcorn Avenue, Toronto, Ontario, Canada M4V 3B2
Penguin Books India (P) Ltd, 11 Community Centre, Panchsheel Park, New Delhi– 110 017, India
Penguin Books (N.Z.) Ltd, Cnr Rosedale and Airborne Roads, Albany, Auckland 1310, New Zealand
Penguin Books (South Africa) (Pty) Ltd, 24 Sturdee Avenue, Rosebank, Johannesburg 2196, South Africa

Penguin Books Ltd, Registered Offices: 80 Strand, London WC2R 0RL, England

Published by Plume, a member of Penguin Group (USA) Inc. Previously published in a Dutton edition.

First Plume Printing, May 2004
10 9 8 7 6 5 4 3 2 1

Ⓟ REGISTERED TRADEMARK—MARCA REGISTRADA

The Library of Congress has catalogued the Dutton edition as follows:
Cutler, Alan.
    The seashell on the mountaintop : a story of science, sainthood, and the humble genius who discovered a new history of the earth / Alan Cutler.
        p.   cm.
    Includes bibliographical references and index.
    ISBN 0-525-94708-6 (hc.)
    ISBN 0-452-28546-1 (pbk.)
    1. Steno, Nicolaus, 1638–1686.   2. Geology—History—17th century.   3. Scientists— Denmark—Biography.   4. Catholic Church—Clergy—Biography.   I. Title.
    QE22.S77C85   2003
    550'.9'032—dc21   2003000125

Printed in the United States of America
Original hardcover design by Erin Benach

To Luz, for all time.

CONTENTS

The Seashell on the
Mountaintop

EDWARD: *Sea-shells, did you say, mother, in the heart of solid rocks, and far inland? There must surely be some mistake in this; at least it appears to me incredible.*

MRS. R: *The history of the shells, my dear, and many other things no less wonderful, is contained in the science called Geology, which treats of the first appearance of rocks, mountains, valleys, lakes, and rivers; and the changes they have undergone, from the Creation and the Deluge, till the present time.*

CHRISTINA: *I always thought that the lakes, mountains, and valleys, had been created from the first by God, and that no further history could be given them.*

MRS. R: *True, my dear; but yet we may without presumption, inquire into what actually took place at the Creation; and, by examining stones and rocks, as we now find them, endeavour to trace what changes they have undergone in the course of ages.*

EDWARD: *This will indeed be romantic and interesting!*

—Granville Penn, *Conversations on Geology* (1828)

Prologue

PILGRIMS

The basilica of San Lorenzo is one of many tourist attractions in Florence. Thousands come each year to admire the pulpits by Donatello, the staircase by Michelangelo, and the sacristy by Brunelleschi, among other splendors. But off to the side of the main church is a small chapel, which is not open to the casual visitor. It is only for the faithful. Inside is a simple altar and a small, limestone sarcophagus. Here, just yards from the crush of sightseers, pilgrims quietly pray. They leave on top of the coffin messages written in many languages, asking its occupant for his benedictions.

The inscription on the wall above the sarcophagus gives his name as Nicolai Stenonis. In the English-speaking world he is known as Nicolaus Steno. The pilgrims venerate him as a saint, but the vast majority of visitors passing through Florence have never heard of him. Not even the most comprehensive guidebooks grant him so much as a footnote.

Yet, during this man's life some three centuries ago, he was a darling of the Medici court and a scientist of unexcelled international stature, a major player in the Scientific Revolution of the seventeenth century. The impact of his genius on the way we today think about the world and our place in the universe rivals even that of Galileo, one of his heroes, who lies buried in the church of Santa Croce, less than a mile away.

He was, first, an anatomist of spectacular skill at a time when the inner workings of the human body were still very much terra incognita. As a young man, he dazzled the scientific world with a string of anatomical discoveries.

But his most revolutionary achievement came afterward, and in a totally different field of science, one that did not really even exist until he had published a slim volume—all of seventy-eight pages, with just a handful of simple diagrams—in which he laid down its blueprint: the science of the earth's geologic history, of deep time.

This was in an era when Catholic and Protestant theologians, so bitterly divided on other issues, were united in their belief that the only reliable source of information about the world's infancy was the book of Genesis. Contemporary scientists, for their part, were still perfecting the "experimental philosophy." To them, the idea that there could be a science of the vanished past was absurd.

But in his little book, Steno showed that recovering the past could be as straightforward as geometry. And in the earth's rocks he had found a testimony no less valid than the Holy Writ itself.

The monuments to part of his story lie outside of the city, beyond Florence, even beyond Italy. These are not marked at all, nor were they created specifically to honor him. Built of upthrust layers of limestone, sandstone, shale, quartzite, and other rocks, they are the earth's mountains. Their pilgrims are scientists, who come to pay their own kind of homage to the man who is, in a sense, their patron saint.

They bring back to their laboratories the same things that medieval pilgrims traditionally brought home as symbols of their spiritual journeys from distant shrines, the same things Nicolaus Steno brought back from the mountains more than three centuries before. They bring back seashells.

SEASHELLS

Nothing lasts long under the same form. I have seen what once was solid earth changed into sea, and lands created out of what once was ocean. Seashells lie far away from ocean's waves, and ancient anchors have been found on mountain tops.

—*OVID*, METAMORPHOSES, *BOOK XV*

The shark was gigantic, but the fishermen managed to haul it ashore. It was still alive and struggling, so to keep it on the beach they lashed it to a tree. Then they killed it. Sharks were common enough off the Tuscan coast, but this was a *lamia*, a great white, and it weighed over a ton. When it was safely dead, several of the fishermen reached into the shark's horrible mouth and with their knives, gouged out teeth for souvenirs and charms.

Word of the marvel reached the Medici palace in Florence. The Grand Duke Ferdinando II, an aficionado of natural history, ordered that the shark be brought at once so that his court scientists could examine it, but it was too huge, and its flesh had already begun to putrefy. The fishermen hacked off the head and threw the rest of the corpse into the sea. The head was loaded onto a cart to be sent up the valley of the Arno to Florence.

The year was 1666. Florence, indeed all of Europe, was in a state of transition. The Renaissance had pretty much run its course. The convulsions of the Protestant Reformation had mostly subsided. The Age of Enlightenment, on the other hand, was barely on the horizon. It was an awkward, in-between age—reborn, reformed, but not yet enlightened.

A generation earlier, the pope had forced Galileo Galilei to renounce his belief in the Copernican theory of the solar system. Galileo had based his opinions on his own observations of the sky rather than on the church-approved texts of Aristotle and the Bible. And though he accepted his punishment, he held firm to his convictions about science. True scientific knowledge, he believed, was grounded in experiment and direct observation of nature, not books, even sacred ones. He lived out his final years in Florence, protected by the grand duke. Now Ferdinando's court was home to a scientific academy founded by several of Galileo's former pupils, determined to keep his spirit alive.

Newest to the group was a diminutive, soft-spoken anatomist from Denmark named Nicolaus Steno. Only twenty-eight years old, he was already famous for his acute powers of observation and his preternatural skill with a scalpel. His discoveries had created sensations in Leiden and in Paris, the twin intellectual capitals of Europe. His bold challenges to conventional theories about the heart and the brain had inevitably made him enemies, but also won him many admirers. The Florentine scientists welcomed him as one of their own. When the monstrous shark head arrived in Florence and was brought into the anatomical theater to be dissected, it naturally fell upon Steno to do the honors.

The chance capture of a shark and its dissection by a young scientist eager to prove himself before a prestigious Italian court mark the unlikely beginning to an intellectual revolution that, in its way, was as profound as that of Galileo and Copernicus. Their revolution had shifted the human position in space: it dislodged us as the fixed center of the cosmos and set our world in motion. Steno's changed our place in time. It removed us from the center of the standard Biblical narrative and gave our world a new history. The time encompassed by this new history expanded from a mere six thousand years to nearly five billion. Vastly older than the human species, the world could no longer be claimed as our exclusive domain.

Steno discovered that the crust of the earth contained an archive of its most ancient history. Up until that time scholars had relied only on the written word, the Bible and the texts of the ancients, to delve into the past. The new philosophers of the burgeoning Scientific Revolution were interested in nature's timeless laws, not its historical development. The chronicle recorded in the earth's geologic strata lay unread, no one truly grasped the stupendous changes the world had undergone over its staggeringly long past. But without this perspective, nothing about the forces that shape our physical world— earthquakes, volcanoes, erosion, climate—could ever make scientific sense. The static, mechanical concept of the world had to be replaced by a dynamic, evolutionary one.

This revolution triggered by Steno in our understanding of the earth gathered momentum slowly; not until the end of the eighteenth century was it in full swing; not until the middle of the twentieth century was it complete. It was resisted as bitterly by scientists as by theologians. It was embraced more readily by Romantic poets than by Enlightenment philosophers. Ironically, the man who launched it never publicly challenged the six-thousand-year biblical timescale that his science eventually overturned. Yet, even in his final years, which he devoted entirely to religion, he never renounced his science, either.

The story of Steno is full of such ironies—and of pathos, as well.

The genius of his ideas was never fully appreciated during his life-time. He died young, at forty-eight. After being a scientist and dar-ling of one of Europe's most lavish courts, he became at the end of his life an ascetic priest. His poverty and fasting, said one friend, had reduced him to "a living corpse."

But that day in Florence he was still at the height of his scientific powers. He had a medical education that, typically for the times, had covered everything from anatomy to astrology. He had the support of a wealthy sovereign. And, most important, he had a mind given to taking unexpected leaps. From the shark it leaped to a seeming unre-lated question, one that, old as it was, could still generate heated de-bate. It was not only his answer to this question, but the way he sought to prove it, that triggered the scientific exploration into the world's distant past.

The question was this: Why are seashells often found far from the sea, sometimes embedded in solid rock at the tops of mountains? The an-cient Greeks had known and written about these seashells. Medieval theologians had noticed them in the building stones of their cathe-drals. Miners and quarrymen found them, as did farmers, shepherds, and travelers. Even the pope in Rome must have noticed them and wondered because they littered the slopes of Vatican Hill.

Today we think it natural to say that the seashells were left by a sea that once covered the land. This, in fact, was the explanation offered by the ancient Greeks. The very earliest of the Greek philosophers, the so-called Pre-Socratics, made it the keystone of their various the-ories of the world six centuries before Christ. Aristotle continued the tradition, writing that the waxing and waning of the seas were part of the world's "vital process." The land naturally experienced many in-undations over the course of time.

Yet most educated people of Steno's time rejected this idea. They thought instead that the shells grew within the earth. Despite all ap-pearances, the seashells were not actually seashells at all. No clams had ever lived inside the fossil clam shells; no seas had ever covered the mountains.

Bizarre as it may seem today, in the context of the time this idea made perfect sense. Some of the more mystical currents of Renais-sance thought were still popular, even among those who prided themselves for their rationality. Neo Platonists and Hermetic philoso-phers had taught that all things on and within the earth were shaped by "plastic forces" and invisible emanations from the stars. No one knew how these mysterious forces and emanations actually produced stones in the shapes of seashells, but the world was a mysterious place: no one knew how a magnet's force caused it to attract an iron bar or orient itself toward the north. No one knew how the sun's "emana-tions" made flowers grow. These things happened in front of the eyes, yet they were still mysterious. Who could say what was or was not possible in the depths of the earth?

The theory that fossil seashells grew right there in the rocks also had the advantage of sidestepping some thorny problems faced by other explanations. There was, for example, a long tradition among Christian writers that fossil seashells were relics of Noah's Flood. The shells were tangible proof of Scripture and a visible reminder of God's power and human sinfulness. Missionaries found them useful for demonstrating to the local pagans that the Flood of the Bible had been universal, not something inflicted on the Hebrews alone.

But a closer look at both Scripture and fossil seashells led to disconcerting questions. There were contradictions, some easier to reconcile than others. The shells resembled species that lived in salt water, but forty days and nights of rain would have made a freshwater Flood. And how could so many shellfish become spread so widely in a flood that, according to the Bible, had lasted no more than a year? Medieval monks had felt free to fudge a little in their reading of the text. Maybe it was the overflowing sea that caused the Flood, not rain, as was written. Maybe it had lasted somewhat longer than the text said. It was a respectable practice. Saint Augustine and the other early Church Fathers had not hesitated to interpret Scripture meta-phorically when necessary.

There was another, stickier, problem, though, one that metaphors couldn't easily solve. The Bible said God had created the solid earth and gave it its form in the very first week. Noah's Flood happened much later. How, then, did seashells get *inside* rocks, which had sup-posedly already been created when the Flood took place? The Flood might have left shells on mountains, but not in them.

Of course, it was possible to call them miracles, and leave it at that, but the budding scientific minds of the seventeenth century were reluctant to do this. They wanted to explain the world by natu-ral laws whenever possible. And since the Reformation, metaphorical interpretations of Scripture had become increasingly frowned upon, too. Luther and Calvin had put the Bible at the center of their faith; the plain meaning of its words was not to be trifled with.

Even the "vital processes" suggested by Aristotle offered no way out of the dilemma. It may have been a perfectly acceptable explana-tion for low-lying shell deposits near the coasts, but for the shells in the mountains, it could lead to dangerous ideas. Aristotle had empha-sized the slowness of his geographical changes. In the time it would take for an ocean to dry up, or a mountain to sink beneath the waves, whole nations might arise and perish. He imagined an eternal world—as many pagans of his time did—which put no limits at all on time, and he claimed that these natural inundations occurred again and again over the ages. For the modern seventeenth-century man, this simply could not be true; there was not enough time. Nothing of the sort had ever been seen over all the centuries of recorded history. Mount Sinai still stood as high as when Moses brought down the Ten Commandments. The Mediterranean Sea had not dried up appreciably, either. How, then, could geography have been overhauled many times when the world itself was known to be less than six thousand years old?

The evidence for this time limit came from the Bible, which was supposed to contain a complete history of the world. By tallying all the generations and reigns of kings recorded in its pages one could estimate the total time elapsed since Creation. The answers varied, depending on which version of Scripture that one used, but none exceeded a few thousand years. The most definitive and precise was the one calculated by James Ussher, the Anglican archbishop of Armagh, Ireland.

Ussher was one of the most formidable scholars of his time; it was said that his personal library was the largest in all of Western Europe. He had devoted his life to compiling his chronology. The date he gave for Creation was Sunday, October 23, 4004 B.C. When he died in 1656, a year after his book was published, the world would have been 5,660 years old, by his reckoning. And he believed, as did many others, that the world was not likely to get much older. Six thousand years would be the limit for the world's total life span.

This was also revealed in the Bible, whose words were assumed to be not only history, but prophecy. The six days of Creation in Genesis foretold that the world would exist for six ages. How long was an "age"? The Bible supposedly revealed this, too. "One day is with the Lord as a thousand years," said Peter, "and a thousand years are as one day." Six thousand years, then, was all the time there would ever be.

For a Christian, the world could not be eternal because only God was eternal. To say that the world was eternal denied that it had a beginning, that it had been created by God, indeed that it had been created at all. People were eternal, too, in the cyclical view of time. This raised all kinds of problems. If a person could borrow money in one cycle, and repay it in the next, which some pagans saw as a perfectly acceptable practice, then where was the urgency for a sinner to reform and repay his debt to God? And the whole idea of salvation was thrown into question if the birth, crucifixion, and resurrection of Christ were not unique events in time, but happened again and again ad infinitum. "God forbid that we should believe this," wrote Saint Augustine. "For Christ died once for our sins, and rising again, dies no more."

The Bible plainly said that the world was created, not eternal. Genesis gave the story of how it happened. Some people, including Saint Augustine, allowed that the six days of Creation were probably metaphorical—although Saint Augustine thought that six days were too *long* for an Almighty God who could create a universe in an instant if he wanted to. But even if one was willing to add a little slack to the creation week, the history of the world still had to be finite. Time went in one direction, it did not loop around, it had a beginning and an end.

Given the difficulty of explaining fossil seashells in a six thousand-year-old world, and the repugnance of the only apparent alternative, Aristotle's eternal world, the idea that they grew in place was understandably attractive. The Flood still had its advocates—Martin Luther notably among them—and a few daring souls even risked openly supporting the eternalist solution. But to argue that the sea-shells actually were seashells was to swim against some strong religious and scientific currents.

This, of course, is what Steno would do. And, along the way, he would offer his ideas on the growth of crystals, the erosion of land, the growth of mountains, and, most famously, the laying down of sedimentary strata. What gelled in his anatomist's mind was a scien-tific approach to the anatomy of the earth, how its parts grew and how their development could be understood. It was, in effect, a new science.

There was no science of the earth's history at the time because the earth was not really considered to even *have* a history. People had a history; not things, not nature. For an orthodox Christian, each part of the world had been created by divine fiat, more or less in its pres-ent form. There was no point in asking how mountains or valleys formed. They had just been created. No further explanation was nec-essary or even possible. If someone allowed that there had, in fact, been a few changes since Creation, these were seen as inherently chaotic. Changes could only mean the decay of God's originally per-fect Creation—changes for the worse, by definition—and so not worthy of Christian contemplation. And, finally, all important events in the six thousand years since the beginning of the world were recorded in the Bible, anyway. There was no need for any further in-vestigation.

Unlike his contemporaries, however, Steno found not chaos, but order in the crust of the earth. It wasn't the perfectly regular order that astronomers found in the heavens, or the mathematical order that physicists found in pendulums and projectiles. It was the order of a well-told story, a narrative in which each part builds on the one before, but the conclusion is not predictable. He found the logical rules by which the faulting, uplift, erosion, and stratification of a landscape and the bedrock beneath it could be put into an intelligible sequence. From the narrative he read in the rocks, he could write a history of the landscape. And the logic, if extended, could reveal the history of the entire world.

The backbone of his system was a simple but tremendously powerful idea. Recognizing that the layers of rock that entombed fossil shells were made by the gradual accumulation of sediment, he realized that each layer embodied a span of time in the past. He saw no way to measure the number of years or centuries involved, and was loathe to speculate, but it was clear that the layers one on top of the other formed an unambiguous sequence. The lowest layer had been formed first, the highest last. Depending on their fossils and their sediments, the layers recorded the succession of seas, rivers, lakes, and soils that once had covered the land. Geologists call Steno's insight the "principle of superposition." It means that, layer by layer, the history of the world is written in stone.

Just as Galileo's telescope had opened up space to science, Steno's strata opened up the past. In an astonishing feat of intellectual focus, Steno produced his seminal geological work in a period of less than two years—two years in which his personal life also underwent major upheavals. Equally remarkable, his new science is outlined in barely one hundred pages of text and just a handful of diagrams. His seventy-eight-page masterpiece *De solido,* "On Solids," was originally intended as an abstract, a "prodromus," of a longer and more detailed dissertation. But that work never materialized. *De solido* was his last published geological work. A few years later he entered the priesthood and gave up scientific research altogether. While in Italy, he had made a controversial switch from Lutheranism to Catholicism, and with the zeal of a new convert he devoted the remainder of his life to the Church.

Steno's singular approach to science made him something of an enigma to his contemporaries. His abrupt retreat from it has made him an enigma to many who have studied him since. He had a mind that was extraordinarily fertile, but also extraordinarily restless. And just as he overturned many assumptions about science, history, and faith cherished in his own time, his story overturns many equally cherished in ours.

CHAOS

Go my Sons, burn your books and buy stout shoes, climb the mountains, search the valleys, the deserts, the sea shores, and the deep recesses of the earth . . . Observe and experiment without ceasing, for in this way and no other will you arrive at a knowledge of the true nature of things.

—PETRUS SEVERINUS,
SIXTEENTH-CENTURY DANISH ALCHEMIST

O n a cold day in March 1659, a slight young man with a pale, angular face, dark eyes, and black hair falling to his shoulders stepped outdoors and looked up at the sky.

It had been a bitterly cold winter, the second in a row. The people of Copenhagen were used to long, dark winters, but the frigidity of the last two years was not only unprecedented—it had brought calamity to the city.

Denmark was at war with Sweden, and the war was not going well. The Danish king, Frederick III, had blundered into the conflict two years earlier, hoping to win territory from Sweden. But the Swedes quickly smashed Frederick's army and immediately overran most of Denmark. Copenhagen was spared only because it lay on the island of Zealand, separated from the mainland by miles of open water. But then came a winter so cold that the straits froze. Thousands of Swedish troops stormed the city. Caught off guard, Copenhagen narrowly avoided destruction.

Now, this winter, Copenhagen was again under siege. There was not enough food in the city and it was hard to find fuel to keep warm. A few weeks before, the Swedes had launched an attack, but the townspeople bravely took up arms and drove them off.

Frail as he appeared, the young man had been there on the barricades just weeks before; he had seen the terror and the bloodshed with his own eyes. The Swedish army still threatened the city.

But right now his mind was far away from all this. He had always amazed his elders with his intense powers of concentration, and now he focused not on the death and misery that surrounded him, but on the pristine flakes of snow drifting down from the clouds. He watched them swirl in the wind, and scrutinized each tiny crystal that lighted on his coat sleeve. He pulled a scrap of paper from his pocket and made rough sketches of their shapes.

The young man was a student at the University of Copenhagen. His name was Niels Stensen, but at the university he followed academic custom and went by a Latin name: Nicolai Stenonis. We know him today by an abbreviated version: Nicolaus Steno.*

That winter, the university had all but shut down because of the war. Steno had entered three years earlier intending to study medicine. He would have preferred to study mathematics, but medicine offered better prospects for a career. The trouble was, his medical studies did not always hold his interest. With classes canceled, he was

---

*Steno published his scientific work under the name *Nicolai* rather than *Nicolaus* Stenonis. His Danish name ("son of Sten") is sometimes spelled *Steensen*; in one surviving signature he wrote his name as *Nicolaus Steensen*. He signed his French correspondence *Nicolas Sténon* and his Italian correspondence *Niccolò Stenone*, adding to the variety.

mostly reading and studying on his own, and too many other subjects beckoned. He kept a journal of his studies, which he entitled *Chaos,* reflecting, perhaps, the jumble of subjects and ideas it contained. Onto its pages he copied his snowflake diagrams, scribbling observations about their perfect, six-sided symmetry.

Such luminaries as Johannes Kepler and René Descartes had written on the shapes of snow crystals. But Steno had been taught not to rely on the authority of books. If he wanted to understand the natural world, he should go and look at it for himself. Sixty years after Galileo, whose observations of the sky had triumphed over Aristotle and Ptolemy, this could still seem a little subversive.

Universities had been founded in the Middle Ages not so much to create *new* knowledge as to preserve *old* knowledge. This meant that knowledge, almost by definition, came from books. Whatever you saw with your own eyes didn't qualify. Even if it were true, it wasn't *knowledge.* Potters learned about clay from the clay itself; miners and quarrymen learned about rocks from the rocks. But scholars, if they had even the mildest curiosity about these materials of the earth, generally contented themselves with what they could glean from the pages of Aristotle or other texts. To dirty one's hands with the things themselves was beyond the pale for academics.

Many universities during Steno's time retained this medieval outlook. Science as we know it today grew up mainly outside of universities. It was the creation of a handful of priests, medical men, courtiers, gentlemen "virtuosos"—as well as a few like-minded professors. Steno would have known it by several names, not all of which meant precisely the same thing; the new philosophy, natural philosophy, mechanical philosophy, experimental philosophy. As the assortment implies, science had yet to crystallize into a unified intellectual pursuit. Older traditions pulled at it, too. Astronomy was not yet wholly distinct from astrology, nor had chemistry separated from alchemy. Even now, in the second half of the seventeenth century, a lot of science still looked like sorcery.

Denmark was a small country, yet it had produced a respectable crop of scientists. Tycho Brahe, an imperious sixteenth-century nobleman with a yen for star gazing, was the most famous. At his observatory on the island of Hven he had mapped the positions of the stars with such unheard-of precision that it revolutionized astronomy. Most astronomers before him, Copernicus included, had hardly bothered to look at the sky. They got their inspiration from books and mathematical tables, not their own desultory observations. The great Tycho paved the way for Galileo and Kepler.

The Danish scientists who followed in his wake were a small, inbred community (most, in fact, were linked by blood or marriage within a single clan, the Bartholin family), but they understood better than most others the lessons of Tycho. Elsewhere, the testimony of authority still overruled that of "experience"—sometimes deservedly so as the enthusiasts for experience were not always scrupulous about the *accuracy* of their observations—but Steno's Denmark was an enclave of empirical science.

For the new philosophers nothing was to be taken for granted; everything must be subjected to the acid test of proof. René Descartes made the "method of doubt" the foundation of his philosophical system. By doubting every proposition until it had been logically proven, all the vague and erroneous ideas of the medieval pedants could be cast aside like so many rotten apples from a basket. What remained would not only be true, it would *necessarily* be true. The new philosophers were determined to explain the world with the same absolute certainty that mathematicians achieved in their proofs and theorems. Dismissing the old Aristotelean or "Peripatetic" style of philosophy still practiced in the univerisities as empty verbal disputation, they took as their model the precise demonstrations of geometry. The "geometric method" would bring together and solidify all knowledge.

"This is the Age wherein all men's Souls are in a kind of fermentation," wrote the Englishman Henry Power, "and the spirit of Wisdom and Learning begins to mount and free itself from those drossie and terrene Impediments wherewith it hath been so long clogg'd."

These are the days that must lay a new Foundation of a more magnificent Philosophy, never to be overthrown: that will Empirically and Sensibly canvass the Phaenomena of Nature, deducing the Causes of things from such Originals in Nature, as we observe are producible by Art, and the infallible demonstration of Mechanicks: and certainly, this is the way, and no other, to build a true and permanent Philosophy.

But if science promised hope and rationality to the elite few who were aware of its progress, it gave no comfort to the vast majority of people. The beauty of snowflakes notwithstanding, Denmark was still at war, and people were still hungry. The benefits of science were still a long way off. Now it offered only change to a world that saw change as dangerous, or, at best, pointless.

Many Europeans were only too aware that the world's allotted six thousand years were almost up. According to Bishop Ussher's calculations, there were still a few centuries to go. But many people agreed with Martin Luther that the world was so wicked that God would probably snuff it out ahead of schedule. The horrible bloodshed of the constant religious wars—that of the Thirty Years' War would not again be equaled until World War I—made the end seem very near. There was not much to do but hold on and wait.

When Steno was born in the winter of 1638, the Thirty Years' War still racked Europe. What began as a revolt of a few Protestant princes in Bohemia against the Holy Roman Empire eventually involved most of northern Europe. Because of the complicated play of religion and politics, at times it nearly degenerated into a war of all against all. By the Peace of Westphalia in 1648, some parts of Germany had been depopulated by more than half.

As if to emphasize the religious rifts of the time, two separate calendars were in use. Depending on the calendar, Steno was born on either January first (Protestant) or eleventh (Catholic). In the preceding century, Pope Gregory XIII had instituted a calendar to correct flaws that had caused the old one to creep slowly out of phase with the solar year on which it was based. The revised Gregorian calendar is now used worldwide, but for more than a century most Protestant countries resisted the change, and not only because they were suspicious of everything that came out of Rome, they didn't think the world would last long enough for the improvements to make a difference.

In solidly Lutheran Copenhagen, the calendar would have read January first the day Steno was born to Anne Nielsdatter and her husband Sten Pedersen, a successful goldsmith. Both parents had been married and widowed before. Steno was the first of two children they had together. A daughter, Anne, was born a couple of years later.

Steno came from a line of rock-ribbed Lutherans, with beliefs that tended toward the bleak. A Lutheran home might have inscriptions on the walls and window frames: "O Man—Remember eternity! God's eye is upon you!" or "Live with death in your thoughts: time hurries, we are but shadows!" Morbid sentiments, but death was everywhere. In 1654, when Steno was sixteen, a plague swept through Copenhagen, killing a third of the city's population. Among his friends and schoolmates, who were enlisted by the authorities to help cart off the bodies, the death toll was even more horrific: In less than a year he saw half of them die. Under the circumstances a strong religious faith, however grim and moralistic, had to have been a comfort.

Steno spent much of his early life isolated from other children. At the age of three he was stricken with an unspecified illness that kept him confined in the family home until he was six. Unable to play outdoors with other children, he learned to pass the time listening to the conversations of his parents and their friends, which usually centered on religion. Not long after Steno recovered from his illness, his father died suddenly.

Sten Pedersen had provided well for his small family. By all accounts he was a highly skilled goldsmith and a good businessman. The Danish king was one of his regular customers, though the king could be casual about paying his bills. Still, the family lived in comfortable quarters above the goldsmith shop in a respectable neighborhood near the university.

But Pedersen's death was a crushing blow to the family's financial fortunes. Anne remarried almost immediately, but this husband, also a goldsmith survived for only about a year. Steno then became the ward of an elder half-sister and her husband, who had a secure job with the government. Eventually, Steno's mother found a fourth husband who was able to keep the family business afloat, but these unsettled early years set a pattern that carried into Steno's adult life. He never owned a home, only once did he live in the same city for as long as three years, and, even at the height of his career, he never had a secure or steady income.

The twin themes of religion and science emerged early in his life. Even when very young, he recalled, he preferred the religious discussions of his elders to "the frivolous chatter of younger companions." In his *Chaos* journal, quotations from religious texts and items of spiritual self-reproach intersperse freely with notes on the latest chemical theories and scribbled ideas for experiments.

Such intimate mingling of science and religion seems strange to us today, but the distance that we now put between the two realms would have seemed equally strange to scientists of Steno's generation. Most of the prime movers of the seventeenth-century scientific revolution were deeply religious. Conflict between true science and true religion was impossible in their minds because both ultimately came from God. Despite his problems with Rome, Galileo remained a devout Catholic until the end of his life. The explicit goal of many scientists, including Robert Boyle, England's doyen of experimentalism and former protégé of none other than the Biblical chronologist James Ussher, was to demonstrate the truth of Christianity through natural law. Newton was no less religious than the others, devoting as much of his energy to Scriptural exegesis and prophesy as to physics.

Science opposed atheism, not religion. Atheists were not allowed to publish books, and the word "atheist" was more a term of slander than anything else, so it is hard to know today what the so-called atheists of the time really did or did not believe. But they were presumed to be, and perhaps many were, impious scoffers and skeptics who denied the existence of universal truths, objective facts, or natural laws of any kind. The stereotypical atheist believed that in nature everything was chance and anything was possible, just as they believed that in human affairs everything was opinion and anything was permissible. Demonstrating the order and regularity of the universe could strike a blow against such dangerous beliefs.

So science would have a perfectly acceptable interest for a pious young Lutheran boy in seventeenth-century Copenhagen. It's unclear what first awakened his interest, but the goldsmith shop just downstairs from the family quarters provided a ready place for a laboratory. There, he ground lenses, mixed chemicals, and built mechanical contraptions. His quick intellect, his pleasant demeanor, and his habit of politely listening to elders—probably even more rare among young geniuses than among the general population of children—made him appealing to mentors. It doesn't take much psychologizing to deduce what the fatherless boy was looking for in these relationships—in later reminiscences he refers to the teachers and older friends who encouraged his scientific interests as his "fathers."

One of these surrogate fathers was an ebullient young Latin teacher and poet named Ole Borch. Borch taught at the Vor Frue Skole, the staunchly Lutheran academy where Steno's family had sent him for the rudiments of a classical education. It was from Borch that Steno learned to write his fluid Latin prose.

Borch's enthusiasms went far beyond literature, however. Originally intending to enter the clergy, he had briefly studied medicine at the university (this was a requirement: rural ministers were often the de facto physicians in their communities), and was something of a local hero because of his care for the sick and dying during the plague of 1654, when most others with medical training had fled the city. But most important for Borch's students, Steno in particular, was the passion these studies aroused in him for the new scientific philosophy. Borch was ultimately to become one of Denmark's most versatile intellects, holding professorships at the university in poetry, philology, chemistry, and botany in addition to a flourishing medical practice. At the Vor Frue Skole, he preached to his students the gospel of nature's book, putting on scientific demonstrations and organizing nature hikes.

It was Borch, more than anyone else, who turned the serious young boy into a serious young scientist. In some ways, teacher and pupil were a study in contrasts—the expansive, poetic Borch; the intense, analytical Steno—but they shared an insatiable curiosity about nature, and both were true believers in the experimental philosophy. The friendship that grew between them continued after both had left Vor Frue; indeed, it lasted until the end of Steno's life.

In November of 1656, at the age of eighteen, Steno enrolled in the University of Copenhagen to study medicine. His timing could hardly have been worse. Within a year, Denmark was at war, and Copenhagen fell under siege a few months later. Many professors and students joined the war effort, others fled to safe havens in the countryside. Fuel to heat the library and other academic buildings became scarce. Class meetings were erratic, many courses were canceled altogether.

Despite the upheaval, Steno managed to get an education—largely, it seems, by dint of his own stubborn efforts and through help from friends, notably the indefatigable Ole Borch. Steno's student journal *Chaos,* written over a period of four months in early 1659, still survives. Mostly it records his eclectic reading—long passages of religious, medical, scientific, and philosophical texts copied verbatim—but there are also fascinating glimpses of his emotional and physical travails, prefiguring both the toughness and sensitivity he would show throughout his life. Oddly, the war itself is never mentioned directly, only its effects on daily life. He complained of the bitter cold and the lack of fuel for a fire. To stay warm enough to study, he did calisthenics. Sometimes it was too cold to do anything.

But most of the anguish that emerges from his personal notes is internal. He berates himself over real or imagined sins of sloth and impiety. Constantly he frets over his own undisciplined mind—its tendency to focus intently on one topic, and then quickly dart to another at the least distraction. Intense periods of study followed by sudden shifts of interest would later characterize his career as scientist, and, indeed, lead to his greatest achievements. But he would always worry about his inability to stick with anything. In *Chaos* he writes: "I pray, thee, O God, take this plague from me and free my soul of all distraction, to work on one thing alone, and to make myself familiar with the tables of medicine alone."

Fortunately for science, Steno's soul continued its wandering habits, despite the earnest prayer. Urged on by Borch, he sampled the main currents of seventeenth-century philosophical and scientific thought, learning the work of Galileo, Bacon, Descartes, Gassendi, Kircher, and others. What the university library lacked—like most university libraries, it was long on classics and theology, but short on science—Borch helped him find in private collections.

Borch also had his own personal laboratory, where he and Steno set aside book learning to conduct their own experiments. Like many of the time, Borch was fascinated by alchemy—not because it offered the prospect of creating gold from lead, but because it appealed to the mystical and poetic side of his personality. Much of alchemical theory was based on the ancient writings of the legendary Egyptian sage Hermes Trimegestus, rediscovered during the Renaissance. Hermetic philosophers saw all knowledge, all reality as a unified whole. To them, it was the book of nature, not so much the book of Scripture, that was to be read metaphorically. They saw signs and "signatures" everywhere in nature. Poetry and love were genuine physical forces in their universe, no barriers precluded one element from transforming into another.

Of course, Borch was limited in the kinds of transformations he could actually accomplish in his laboratory. But, as with others who dabbled in the alchemical arts, he did gain from his efforts some practical knowledge of chemical reactions. Steno, for whom alchemy never held much appeal, was nonetheless deeply impressed by some of Borch's experiments. Years later, at the start of his geological studies, he would think back to the things he had seen in his old teacher's laboratory—stones dissolving in water, crystals growing out of clear liquids. He wrote of one experiment in which Borch caused particles of "earth" to precipitate from a limpid fluid. What he had observed happening in Borch's laboratory he could imagine happening in an ancient sea: grains of earth settling slowly through the water, accumulating on the bottom in a deepening pile. When he wrote about the sedimentary deposits he envisioned, he would describe their arrangement by an expression borrowed from the laboratory: *stratum super stratum,* layer upon layer.

Steno was at the university to study medicine, not the nonexistent field of geology, however. His preceptor, the professor in charge of overseeing his studies, was Thomas Bartholin, head of the university's three-man medical faculty and a scion of the academically powerful Bartholin family. Famous for being the first to discover lymph vessels in the human body, Bartholin was Denmark's leading anatomist.

Steno's interest in anatomy had probably been whetted by Bartholin's popular presentations at Copenhagen's anatomical theater. In seventeenth-century Denmark, anatomy was a prestige science: the physical details of human anatomy were as new and exciting as the

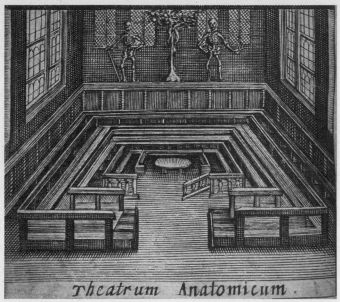

Copenhagen's Anatomical Theater, 1662. Human skeletons "Adam" and "Eve," flank the Tree of Knowledge with the serpent in its branches. Thomas Bartholin, *Cista medica hafniensis* (1662). Courtesy of History of Science Collections, University of Oklahoma Libraries.

sequences of the human genome are today. The theater was the forum where university researchers shared with the gaping public their latest discoveries about the human body—usually one fresh from the gallows. Bartholin was especially known as a crowd pleaser, intoning eloquently as he removed each organ from the corpse to pass among the audience, which might include the king and members of his court. Ticket sales covered instruments, hangman's fees, and refreshments, plus salaries for Bartholin and his assistant, who made extra money on the side by making belts from cadavers' skins and selling them to unsqueamish patrons.

As it happened, Bartholin retired from teaching just weeks after Steno entered the university, and he was absent from Copenhagen for long periods during the war. But somehow Steno and Bartholin managed to forge some kind of bond, informal and sporadic though it must have been. Bartholin was the man who launched Steno on his first scientific career—anatomy. And it was most likely Bartholin who first introduced Steno to the perplexing question of seashells on mountaintops.

Mountainless Denmark afforded no opportunities to observe such things, but Bartholin was well-traveled. It was customary for gentleman scholars to round out their university educations with a "Grand Tour" of the continent. Bartholin's tour lasted ten years. During this time he ranged widely in Europe, but mostly he stayed in mountainous and fossil-rich Italy, where fossil seashells were a perennial subject of speculation. He had even made a special trip to the island of Malta to observe its famous fossil shells and rich deposits of medicinal glossopetrae, or tongue-stones.

Bartholin's interest in the fossil question was strictly professional. It was a physician's business to know about stones of all kinds. Crystals, mineral ores, fossils, and assorted other earthly objects were generally believed to have curative powers. The iron-ore hematite, for example, could ward off diseases of the blood. Crystals of topaz, if pressed against a wound, could staunch bleeding or, if worn as an amulet, prevent nightmares. Carved into the shape of a dragon, the stone jasper soothed an upset stomach.

He had a collection of fossils from his journeys, which he showed to Steno. He was even preparing a book on the subject of their medicinal virtues, along the lines of one he had already written on unicorn horns. Unicorn horns had long been considered to be potent cures for any number of ailments, and Bartholin's book had greatly impressed the medical community of the time, despite the facts that unicorns were mythological creatures and the horns in question had already been revealed to be narwhal tusks. But that wasn't important to Bartholin so long as the medicine worked. Similarly, he was not particularly interested in the debate over the origin of fossils; he mainly cared about how to use their curative powers to best advantage. If pressed he allowed that fossils probably grew in the earth, sprouting from the ground "like a plantation" of stones.

The debate over the origin of fossils had always been complicated by a certain confusion over just what was meant by the word "fossil." Traditionally, it referred to *any* natural object or substance dug from the earth. The modern term "fossil fuel" is a linguistic relic from this time. Today the word is generally used for the preserved remains of ancient plant or animal life—bones, teeth, shells, wood, and so on— found in rock strata, but it formerly included all distinctive stones, crystals, gems, and mineral ores that one might dig up. Most of the objects illustrated in *On Fossil Objects,* published by the German naturalist Conrad Gesner in 1565 and the most authoritative work on fossils until Steno's time, would not be called fossils today. One woodcut in the book depicts a wooden cylinder holding a piece of lead, a metal dug from the earth. The accompanying text explains the usefulness of this strange device, now known as a pencil. Fossil stones resembling parts of animals or plants were called by a variety of terms, such as "figured stones" or "extraneous fossils."

Steno left no clue as to where he stood on the thorny question of "extraneous" fossils at this stage in his life. His notes in *Chaos* contain numerous references to fossils in the broad sense, mainly copied word for word from medical texts without comment. Many are simply advice for the practicing physician on how to use which stone for which ailment. But the student's curiosity was obviously piqued by the possible biological connection—and by its implications for the world at large. In one passage copied into *Chaos* he underlined the last words for emphasis:

> *Snails, shells, oysters, fish, etc. found petrified on places far remote from the sea. Either they have remained there after an ancient flood or because the <u>the bed of the seas has slowly been changed.</u>*

THE ANATOMIST

Why, then the world's mine oyster.
Which I with sword will open.

—WILLIAM SHAKESPEARE,
THE MERRY WIVES OF WINDSOR

"This M. Steno is the rage here," wrote a physician visiting Paris in the spring of 1665. "This evening after dinner we saw [him dissect] the eye of a horse. To tell you the truth, we are only apprentices next to him. I begged him to show me a heart tomorrow morning which, with singular good will, he promised to do. He is constantly dissecting. He has patience that is inconceivable, and with practice he has acquired a technique above the ordinary. Neither a butterfly nor a fly escapes his skill. He would count the bones of a flea—if fleas have bones."

The physician was one of many who were impressed by the talents of the young man from Denmark. After one of his public dissections at the Sorbonne's prestigious École de Médecine, a Parisian journal wrote: "he makes most of what he presents so vivid that one is obliged to be convinced, and one may only wonder that it has escaped the notice of all earlier anatomists."

The goldsmith's son had obviously inherited some of his father's manual skills. He was staying in Paris at the home of Melchisédec Thévenot, a wealthy diplomat and patron of science. The social and scholarly elite came by to pay their respects and to watch him perform. They were not disappointed. "I have never seen such dexterity," the French physician wrote again, two weeks later. "He made us see everything there is to see in the construction of the eye—without putting either the eye, the scissors, or his one other small instrument anywhere but in his one hand, which he kept constantly exposed to the gathered company."

But it wasn't just his parlor tricks that captured everyone's attention. Behind the pleasant, unassuming manner was a mind deft and sharp as his scalpel. Not long after arriving in Paris he had lectured on the anatomy of the human brain. In perfect French, he presented the theories of the leading authorities—Galen, Willis, Descartes—and then calmly, with unsettling ease, cut them all to ribbons. A book he had written on the structure of the heart and muscles threatened, according to one reviewer, to "turn upside down what is basic in medicine."

Steno had become famous. Not six years before he had scrounged for books and firewood in Copenhagen's half-deserted university. Now his own books were read and discussed by scholars across Europe. He was an honored guest in the best salons, mingling with the most brilliant and sophisticated people.

Steno had not managed to earn a degree in Copenhagen, despite his feverish studies. Perhaps the situation was too chaotic. Instead, he scraped together funds and in the fall of 1659 left Copenhagen, somehow avoiding the Swedish troops that still encircled the city.

Armed with a letter of introduction from Bartholin, Steno meandered across northern Germany for several months, dropping in on his teacher's friends for lodging and informal medical instruction. By springtime he had arrived in the Netherlands, where he resumed his studies in earnest.

His first stop was Amsterdam, then the richest and most vibrant city in Europe. With a population of one hundred fifty thousand, five times that of Copenhagen, it had largely escaped the conflagrations of the Thirty Years' War. Now it was cashing in handsomely on the swelling Atlantic trade. Exotic scents wafted from the brick warehouses of the East India Company and a din of languages rose from the streets: Dutch, Flemish, English, Arabic, Hebrew, Spanish, Italian, and others. Books on every imaginable subject fairly spilled from bookstores. And, unlike in Denmark, where Lutheranism was universal and strictly enforced, the people openly practiced a diversity of faiths. Calvinists, Catholics, Anglicans, Lutherans, Jews, and Muslims brushed elbows on the way to their houses of worship.

For a young man hungry for new experiences and new ideas, no place on earth could have been more thrilling; Steno wasted no time before sampling its riches. But for one accustomed to religious certainty and ever anxious for his soul, no place could have been more disquieting. The cascade of doubts that began in Holland would later culminate in an all-consuming religious crisis.

He stayed in Amsterdam just three months, moving on to the university in Leiden. Before leaving, he wrote a short thesis on the subject of hot springs—his first scholarly publication. The topic today falls solidly within the field of geology, but then would have been a standard part of a medical education. Hot springs and mineral waters were greatly valued by physicians for their healing effects. "*De thermis*," "On Hot Springs," gives few hints of his later geological ideas. It is no more than a preliminary overview of the subject; it reads like a classroom exercise, which perhaps it was. If Steno was already thinking about scientifically probing the earth's structure, these ideas were quickly swept away by a chain of events set in motion within weeks of his setting foot in Holland.

What happened was that, quite unexpectedly, he made his first scientific discovery. And, equally unexpectedly, he became the center of a bitter personal controversy. It was a baptism by fire into the rough-and-tumble world of academic politics.

Steno's host in Amsterdam was Gerard Blaes, the city physician and an old friend of Bartholin. Blaes regularly gave public lectures on medical topics, which he invited Steno to attend. He also offered Steno private lessons on anatomy.

One afternoon, Steno bought a sheep's head from a butcher shop, and brought it back to Blaes's laboratory, intending to dissect the brain. Before opening the skull, he first explored some of the more accessible parts.

"I happened to decide to investigate first the course of the veins and arteries at the mouth by introducing a probe into the vessels," he wrote. Then, "I suddenly discovered that the point of the probe was moving freely in a spacious cavity and struck with ringing sound against the teeth. Surprised at this, I called my host to hear his opinion."

Blaes was not impressed by Steno's find—not at first, anyway. He said Steno had pushed too hard and punctured the tissue with his probe. When Steno showed that wasn't the case, Blaes dismissed the aperture as a "freak of nature." There was no mention of the tiny duct in any of the anatomy texts Blaes had in his library. The subject was dropped. Summer came and Steno left for Leiden to enroll at the university.

Steno repeated the dissection in Leiden, showing the duct to his anatomy professors, who immediately confirmed that it was new to science. He had found the duct of the parotid gland, which supplies saliva to the mouth. Up to that point, neither the source of saliva nor the function of the gland was known. It was by no means a major discovery, but certainly a respectable accomplishment for a novice. Generations of anatomists before him had completely missed it. The professors gave a public presentation of the student's discovery, naming it the *ductus Stenonianus,* Stensen's duct, the name it goes by today.

In Amsterdam, Blaes caught wind of the attention Steno and his saliva duct were getting. Deciding that it was he, not Steno, who deserved credit for the find, he sent letters to the Leiden professors blasting the "wretched boy" with every epithet he could think of: deceit, ingratitude, bad manners, blundering, foolishness, perfidy, incivility, treachery, calumny, scoffing, arrogance, perversity, shamelessness, impudence, and depravity. He rallied comrades in Amsterdam to join the attack. Steno was denounced publicly as a fraud and an intellectual thief. Barely a year in the Netherlands, the student found himself the object of an organized smear campaign. Scientists as far away as Italy heard about the fracas.

For the rest of his career, Steno would refuse to be drawn into disputes over credit for discoveries, a scruple that ultimately cost him some of the fame he deserved. But this time he fought back. Not only was it his first discovery, but his integrity had been impugned. He had to defend his reputation.

Blaes followed his initial volley of invective with a published account of his "discovery," adding to Steno's consternation. Not only was the bogus claim now in print, but Blaes hadn't even bothered to get the anatomical facts right. "If I were not certain that I had shown him the duct," wrote Steno to Bartholin, "I might allege that he had never seen it."

Steno's battle with Blaes went on for months, each side assailing the other in public lectures and printed pamphlets. On the face of it, the match was uneven. Blaes was an established physician, with a solid, though not brilliant, reputation. Steno was an unknown, an upstart. Even some of his friends thought his counterattacks were unseemly. But Steno did have an advantage: whatever polemics Blaes and his partisans might hurl at him on the printed page, none had his delicate touch with a knife. At the dissecting table Steno could show all comers that it was his version of the duct's anatomy that matched reality.

And to prove that the first discovery was no fluke, he launched into a frenzy of dissection, intent on making more. He delved into delicate tissues that clumsier hands would have turned into hash. Within a year he had a manuscript describing in detail not just the salivary glands, but all the glands in the head. He discovered, for example, the tear-producing glands. Tears were previously assumed to be literally squeezed out of the brain by the force of grief.

If the loud accusations from Blaes first brought Steno to public notice, the undeniable quality of his work now kept him there. "Anatomical Observations on Glands," published in early 1662, was a tour de force. Ole Borch, visiting Steno in Leiden, wrote to Bartholin: "In truth he is a genius, worthy of growing into his country's hope."

By the time of his debut in Paris three years later, Steno had expanded his investigations far beyond the lowly glands. He had also developed some strong ideas about how science should be practiced.

Steno was unusual among his colleagues not only for his skill at the dissecting table, but for the fact that he dissected at all. Most anatomists were unwilling to bloody their own hands and left the work to an assistant. In fact, at most medical schools dissection was more like an academic ritual than a method of scientific research.

The ancient texts of the Greek physician Galen had been the primary source of anatomical knowledge for nearly fifteen hundred years. Galen's authority often trumped that of actual cadavers. Dissection was the art of opening up flesh to reveal what Galen said was supposed to be there. If what was found did not match the text, it was an embarrassment to the dissector, not to Galen.

Stubborn adherence to tradition was to Steno the chief obstacle to progress in science, especially anatomy. Incisions were made according to prescribed rules, and the organs were to be examined in a prescribed order. Such rigidity was completely contrary to genuine scientific research, said Steno, which, "does not admit of any set method, but must be attempted in every way possible."

Sometime in the winter of 1665, shortly after he had arrived in Paris, Steno presented a lecture on the brain at the home of his host, Melchisédec Thévenot. Besides his geological treatise *De solido, Discourse on the Anatomy of the Brain* is his best-remembered scientific work. It was not so much an anatomy lesson as a manifesto on his evolving philosophy of science.

He startled the assembled savants by declaring at the outset: "Instead of promising to satisfy your inquiring minds about the anatomy of the brain, I confess to you here, honestly and frankly, that I know nothing about it." It was not false modesty. He had come to the conclusion that everything previously written about the brain was so inaccurate and contradictory that it was better to admit ignorance and start from scratch.

Part of the problem stemmed from the delicate tissue of the brain itself. "Every anatomist who has been concerned with dissecting the brain can demonstrate everything he says about it," he said. "Because its substance is soft and so compliant that his hands, without his thinking about it, shape the parts as he envisaged them beforehand."

As much as scientists wanted to believe that their observations were unbiased and their conclusions certain, the same question could yield different answers depending on how it was asked. It was no wonder that centuries of dissection by the same traditional methods had always given the same traditional results. Many scientific debates went unresolved because the two sides did not use the same method of study.

He went on: all too rarely did authors clearly distinguish fact from speculation in their writings. Instead, they presented their theoretical systems as truths. How could science possibly build on such unsolid foundations?

A particular target was René Descartes, who claimed to have solved the centuries-old quest for the seat of the human soul.

Descartes had been dead for fifteen years, but his posthumously-published book *On Man* had just been translated into French, and was the talk of Paris. In typical iconoclastic fashion, he argued that the human body was simply a machine made of dead matter; all life processes and every action of the body could be explained entirely by mechanical principles.

In the mechanical philosophy nothing was inherently alive, not even plants or animals. To Descartes, a dog was just a machine; its movements were as mechanical as the movements of a clock. If it barked or yelped in pain, he heard it as the screeching gears or ringing bells of an automaton.

As a scientific proposition, the mechanical philosophy offered the exciting possibility that the material world was essentially a simple place. The machines that made it up might be horrendously complicated, but they were built from simple parts that worked in simple ways. Through science and reason the human mind would eventually be able to understand everything.

This picture of the world left out one important thing, however: the soul. In other philosophies souls were everywhere, but in the mechanical philosophy they were nowhere. Descartes denied that animals had souls, so they weren't a problem. But humans surely had souls. How and where were soul and body connected? How could the mechanical philosophy account for the effect of an immaterial soul on a material body?

In *On Man,* Descartes declared the pineal gland, a small nutshaped gland in the center of the brain, to be the crucial link. Twisting and turning in response to the soul's demands, it literally pulled the strings that controlled the body's movements.

Since his student days in Copenhagen, Steno had greatly admired Descartes. In Holland he had become friends with the Jewish philosopher and erstwhile lens-grinder Baruch Spinoza, who was writing a book on Cartesian philosophy. The mechanical approach to anatomy and Descartes's doctrine of doubting all facts and propositions until definitely proved—these both appealed to Steno's analytical mind.

But on the key issue of the soul, the great Descartes had stumbled badly. Supremely confident in his powers of deduction, he had hardly bothered to look at any actual brains to confirm his theory. The little Descartes knew about the pineal gland had been based on a few sloppy dissections. But Steno had found, and demonstrated to his audience, that in an undamaged brain the pineal gland was immobile, held fast by tissues that rough handling inevitably tore. There was no way the gland could perform the gyrations that Descartes's theory required.

Steno's public refutation of Descartes added fuel to what was already a bitter controversy over the dead philosopher's ideas. To Steno's dismay, many of Descartes's supporters simply refused to accept the visual evidence of Steno's dissections. They held on to the teachings of their master as stubbornly and dogmatically as the Galenists ever did. On the other side, religious enemies of Descartes, seeing nothing but heresy and atheism behind his mechanical philosophy, tried to enlist Steno's science in their efforts to suppress his work altogether.

As for Steno, the failings of Cartesian science shook some his earlier confidence in the reliability of human reason, and the unresolved questions of the body and soul added to some of the spiritual anxiety that had been gnawing at him since Holland. But he remained steadfastly committed to the mechanical approach to science. The mistake of Descartes was to put too much faith in purely deductive reasoning for his picture of the world, and too little in empirical observation of the real world around him. He failed to subject his own ideas to the same scrutiny of doubt that he did those of others. Later in his life, after leaving science, Steno wrote: "I do not reproach Descartes for his method, but for ignoring it."

In addition to glands and the brain, Steno had become fascinated by muscles, which for the mechanical philosophy posed almost as big a paradox as the seat of the soul. In the mechanical view, all movement came from the push of one solid object against another. No object could generate motion on its own. How, then, could a muscle contract? What "pushed" it? Steno thought that the question needed to be approached "geometrically." He spent the summer of 1665 at Thévenot's country estate with a friend from Leiden, Jan Swammerdam, dissecting a variety of animals. A moody and socially inept loner, Swammerdam was even more intense than Steno and had made his career dissecting insects under the microscope. Meantime, Steno was deciding what to do next with his life.

Five years earlier a chance discovery had launched Steno's career. Since then, science had won him glory and the respect of Europe's finest minds. It had not, however, bought him security. Both his mother and stepfather had died; the family business was in the hands of his sister and brother-in-law. His own small inheritance was dwindling. Two potential jobs in Copenhagen had fallen through. Though he had earned an M.D. from Leiden, he was not interested in practicing medicine—he had come to the conclusion that most of the traditional cures for disease were worse than useless. The mechanical approach to anatomy would someday provide scientific remedies for the body's ailments, but that day was still far off so far as he could tell.

For the time being he was a welcome guest in Paris. There was talk that Louis XIV might charter a French academy of science—King Charles II in England had chartered the Royal Society some three years earlier—and Steno's friends intimated that if he stayed in Paris he would be a strong candidate for membership.

But Steno was restless. Perhaps he had already set his sights on joining another group of scientists, disciples of Galileo dedicated to the experimental method, who had formed an academy in Florence almost ten years earlier. Whatever his plans, in the fall of 1665, after less than a year in Paris, Steno packed his bags and set off on a new journey.

As an anatomist he had learned to see beneath the surface of the human body. He had dispelled the naive myth that science could provide quick answers to the deepest questions: The seat of the soul could not be found by slicing open heads, or by pure deduction either. It was not a message everyone wanted to hear.

Yet he had also learned that the subtlest details—a tiny pore inside the cheek, the arrangement of fibers in a muscle—often led to unexpected discoveries, sometimes to profound new insights. Each layer of flesh that was peeled away told something different about the life of the body. It was not so much in the dissector's blade as in his eye.

On his way to Florence, Steno traveled through the south of France, then through the Alps and Apennines over a period of six months with many stops and side trips. If he had never before had the chance to see with his own eyes layers of rock packed with fossilized shells, or strata raised and contorted into mountains, he certainly saw them then.

A mountain cannot be opened up with a scalpel; the earth's strata cannot be cut away to see what lies below the surface. But in Florence, Steno would learn to do that figuratively, if not literally. He would become an anatomist of the world.

THE TESTIMONY OF THINGS

Since things are far more ancient than letters, it is not to be wondered at if in our days there exists no record of how the aforesaid seas extended over so many countries . . . But sufficient for us is the testimony of things produced in the salt waters and now found again in the high mountains, sometimes at a distance from the seas.

—LEONARDO DA VINCI, *NOTEBOOKS* (1508–1518)

Some two hundred million years ago, during the span of geological time called the Triassic Period*, the region of the earth's surface now occupied by the Italian peninsula was covered by a warm, shallow sea. There was no hint of today's mountainous landscape, just thick deposits of limey muck, left on the sea bottom by innumerable generations of mollusks, corals, plankton, and marine algae. When later shifts of tectonic plates caused the crust here to buckle, the intense pressure and heat transformed the soft marine ooze into hard rock, creating in some layers a fine-grained and dazzlingly pure white marble, which, like snow, capped the mountains rising from the sea.

Out of this stone, quarried in Carrara and other mountain towns

---

*See page 205

of the rugged Apuan Alps, Italian sculptors created some of the world's great treasures of art. In the best marble, the former sediments have been so completely recrystallized and homogenized that no traces of the original marine creatures are visible. The images artists claimed to find within the rock came from their own imaginations. In blocks of Carrara marble Michelangelo saw his *Moses,* his *Pietà,* and his *David.* A child of the sandstone quarries near Florence, he knew his rocks. Michelangelo climbed the mountains searching for outcrops of the purest marble, which he claimed for his stonecutters by carving the letter "M" onto the rock face.

Also roaming the Italian hills around this same time was another artist, Michelangelo's arch-rival Leonardo da Vinci. Leonardo was less interested in the marbles and hard sandstones of Michelangelo's quarries, however, than in the softer layers of sand and clay. Here, he discovered images that were not creations of his imagination, though they surely aroused its formidable powers. The hills were full of oysters, clams, cockles, and other seashells.

Long after the Carrara marble had hardened to stone and the crust began to crumple and build the mountainous spine of Italy, the sea had reinvaded the flanks of the peninsula. The deposits left behind were carved by rivers into ridges and valleys, but were otherwise undisturbed. Leonardo grew up in the town of Vinci, nestled among these marine hills. As the illegitimate son of a minor public official his formal education was minimal; he must have observed the layers of sand and shells and drawn his own conclusions well before he encountered the conventional explanations in books.

Of these explanations, one that he found particularly dubious was that the shells were relics of Noah's Flood. As the archetypal Renaissance man, Leonardo was naturally conversant with the habits of mollusks. He knew that even the speediest clam could only creep a few feet a day. "At such a rate of motion," he wrote in his notebook, "it would not have traveled from the Adriatic Sea as far as Monferrato in Lombardy, a distance of 250 miles, in forty days—as he has said who kept a record of this time."

Leonardo made other observations: the shells varied in size and species; some were clamped shut as if buried alive on the sea bottom, others were scattered and broken as on a modern-day beach. On some shells he counted yearly growth bands. Though petrified and embedded in rock, the shells sported chips, fractures, wormholes, encrustations, and other marks of life on the seafloor. In the layers of indurated sediment, he discerned the meandering tracks of worms.

The testimony of the shells was clear: They had lived normal molluscan lives and had gradually accumulated over an extended period of time. It was the sea, not a catastrophic flood, that had put seashells in the hills.

Leonardo wrote extensively on fossils, no one else of his era made such acute observations or drew such insightful conclusions. But he never published any of it. It was all to go into a massive "Treatise on Water"—one of his many unfinished projects.

So when Steno made his way through the same Italian hills during the summer of 1666, perhaps indulging his curiosity now and then to pick up a rock or shell from their slopes, he would not have known anything of Leonardo's ideas.

But arriving in Florence during the hot month of July, he was there just in time to witness one of the most famous experiments in the history of science. It cannot have been a pretty sight. The odor must have been abominable.

The scientists in Ferdinando d'Medici's court proclaimed their devotion to the experimental method in their motto *provando e riprovando,* meaning "test and retest" or "prove and refute." They probably had lifted it from Dante's *Paradiso*:

> Quel sol che pria d'amor mi scaldò il petto
> di bella verità m'avea scoverto
> provando e provando, il dolce aspetto.
> *That sun which first inflamed my breast with love*
> *Uncovered for me, with proof and refutation,*
> *The sweet-shining features of the lovely truth.*

But the bright Tuscan sun that warmed Dante's heart must have raised an unholy stench for Francesco Redi, the grand duke's physician, when he set out lumps of ox dung and a smorgasbord of rotten meat to slowly putrefy in the heat. The idea was to test the ancient doctrine of spontaneous generation. According to this view, the process of decay could bring to life lowly creatures such as flies and "an infinite number of worms."

True to the spirit of testing and retesting, Redi had spared no effort in his study, trying out the flesh of every creature he could lay hold of: fish, frogs, snakes, sparrows, pigeons, swallows, chickens, ducks, geese, goats, lambs, rabbits, deer, dogs, oxen, horses—even a tiger and a llama from the ducal menagerie. Some meat he cooked, some he left raw. Some he left to rot in the open air, some he covered with veils to keep out the ubiquitous summer flies.

Athanasius Kircher, a Jesuit scholar in Rome, claimed to be able to generate butterflies, bees, scorpions, frogs, and snakes at will from various blends of excrement and putrefied flesh. He had even published recipes.

But Redi, no matter what he did, only got flies, the very same kinds of flies, in fact, that buzzed everywhere in the palace. What's more, they only appeared when he left the meat or dung uncovered. This meant, he reasoned, that the flies came from eggs laid by earlier flies. Whatever the appearances, putrid matter was just a breeding site for vermin that swarmed over it; it did not produce them. Neither dung nor rotten meat had an intrinsic power to generate life.

It was a simple experiment, but it made its point. History remembers it as the first "controlled" experiment (because of the fly-proof veils), and the first real scientific attempt to refute the doctrine of spontaneous generation.

Redi was one of the people Steno had come to Florence to meet. Twelve years older than Steno, witty and debonair, he was a consummate creature of the Baroque court. With impeccable style he had shielded the pieces of meat and dung in the fly experiments with "a fine Naples veil." Equally at home in the literary salon as in the laboratory, Redi was perhaps more appreciated by his contemporaries for his bawdy poetic tribute to the wines of Tuscany than for his scientific achievements. *Bacchus in Tuscany,* which he composed primarily for the amusement of himself and his friends, is now considered a classic of seventeenth-century Italian verse.

Redi also had the honor of being a charter member of the famous *Accademia del Cimento,* "Academy of Experiments," the world's first scholarly academy wholly dedicated to experimental science. Counting among its ten members two of Galileo's most brilliant students, Vincenzio Viviani and Giovanni Alfonso Borelli, the Cimento was to carry on his battle against the stale Aristotelianism of the universities. It was the brainchild of the grand duke himself, Ferdinando II d'Medici, and his younger brother, Prince Leopoldo.

In setting up their hothouse for the new experimental philosophy, the Medici brothers were following an old family tradition. Wielding power off and on for nearly four centuries, the Medici clan had always been patrons to the greatest artists and philosophers in Florence. Medici money paid for the art of Michelangelo and Botticelli; it underwrote Marsilio Ficino's translations of Plato. Niccolò Machiavelli had dedicated *The Prince* to a Medici patriarch, Lorenzo the Magnificent.

But Ferdinando and Leopoldo were hardly Machiavellian types, nor could either be described as magnificent. Those genes had disappeared from the bloodline. Almost comically ugly, but intelligent and open-minded, Ferdinando had no stomach for politics or, it seems, any of his official duties as the Tuscan ruler. He preferred philosophy, particularly science, which he had learned at the knee of his old tutor, Galileo. When he could sneak way from affairs of state, and from the grand duchess, who was much alarmed by his modern ideas, the gentle Ferdinando tinkered with thermometers and other scientific instruments, a pastime that suited him far better than ruling a country.

Prince Leopoldo shared his brother's affability and intellectual bent. He bought artwork and books by the crate. He read everything: natural history, theology, philosophy, sending his agents abroad to find "books of criticism, gallantry, satire and curiosities," "tall tales with a bit of spice—anything will do." "Like boys with a piece of bread," wrote the Cimento's secretary Lorenzo Magalotti, "he always keeps a book in his pocket to chew on when he has a free moment."

Leopoldo, more than Ferdinando, oversaw the Cimento's operation and set its agenda. He built laboratories in the Pitti Palace and commissioned craftsmen to manufacture beautifully ornate glassware and measuring instruments for the academy's experiments. Occasionally running experiments of his own, Leopoldo loved nothing more than to mingle with the scientists in the lab and to be included in their discussions "as an academician, not as a prince." With his tact and good humor, he kept the scientists' unruly egos in line—it was understood they were only to speak kindly of one another in his presence—and, above all, he kept the money flowing.

Nowhere else in Europe did a monarch take such an interest in scientific research or support it so lavishly. Certainly no university was nearly so hospitable to the subversive new experimental philosophy. The Cimento was unique. In their wigs and brocaded jackets, Ferdinando and Leopoldo's experimenters dissolved pearls in acid and used plates of solid gold to prove the penetrating force of magnetism, all to the delight of the court.

For someone of Steno's interests it was close to paradise. Corpses could be had from the hospital or gallows with a wave of the grand duke's hand. The court menagerie supplied a wonderful variety of exotic animals to dissect. And should one desire a particular kind of beast not already on hand, His Serene Highness would dispatch hunters or have his agents abroad obtain specimens. A naturalist could follow his curiosity to Medici villas in the mountains or by the sea, or browse collections in the palace museum, easily one of the finest in Italy.

And Steno found that, contrary to the stereotypes of Catholics drilled into him in Lutheran Denmark, his new friends turned out to be devout, morally upright Christians. The carnivals, feasts, and rituals of Catholics and the opulence of court life, even the endless bottles of Redi's beloved Tuscan wine, had paradoxically left their religious zeal undiminished.

Looming over everything was the ghost of Galileo. Both Medici brothers had studied with him, as had several of the academicians. Those old enough to have known the master personally often reminisced about their singularly close friendships with him. The mathematician Vincenzio Viviani, Galileo's biographer and very last pupil, turned his home into a veritable shrine to the revered master, arranging that at his own death he would be buried alongside Galileo in his crypt.

The issue of Galileo's condemnation by the church was a delicate subject, to say the least. Most attributed the unfortunate events to the personal jealousy and vindictiveness of enemies among conservative Aristotlean scholars. Galileo was naturally innocent of any indiscretion or impiety, and in no way was his science contrary to faith.

Still, the ban on teaching Galilean astronomy remained in effect, even if it could be easily gotten around by calling the sun-centered system a "hypothesis." Church censors still scrutinized everything to be published for ideas contrary to orthodoxy. With his princely status, Leopoldo might blithely read prohibited books with impunity, but ordinary academicians had to be more careful. They stuck to measuring temperatures, air pressures, and the periods of pendulums, and stayed away from risky metaphysical speculation. The old Aristotelean order would have to be replaced empirically, test by test, fact by fact.

It was just this kind of careful, objective approach to experimentation, however, that had attracted Steno to Florence. Everywhere else he had been disappointed by the dogmatism and self-indulgent speculation that too often passed for science. Florence offered no end of new opportunities. One of which was soon delivered to Steno, practically on his doorstep, in the form of a giant shark.

TONGUES OF STONE

Facts which at first seem improbable will, even on scant explanation, drop the cloak which has hidden them and stand forth in naked and simple beauty.

—GALILEO GALILEI,
DIALOGS CONCERNING TWO NEW SCIENCES (1638)

When the shark that launched Steno's geological career was spotted by a French fishing boat in October 1666, Steno had spent the summer working on a theory of muscle contraction that he based, in a true Galilean fashion and with the help of Viviani, on the firm principles of geometry. The grand duke was footing the bill to have it published. The manuscript had just been approved by the Church censors when word came from Livorno that the shark had been captured.

Ordinarily, the grand duke would have probably directed Redi to handle the dissection. He was an able dissector, with a theatricality and poetic flair that made his scientific demonstrations popular with the other courtiers. But this time he was a spectator only. Ferdinando chose Steno.

Dragged onto shore and clubbed to death, the shark, a great white, tipped the scales at 3,500 Florentine pounds—about 2,800 pounds in today's measure. By the grand duke's order it was butchered, the body and entrails cast back into the sea, and the head sent on its way to Florence, where Steno waited.

When it arrived, a crowd of scientists and other courtiers, probably the grand duke and Leopoldo as well, gathered to watch Steno perform the dissection. The huge bloody head of this monster from the deep, with its bulging eyes and its maw bristling with wickedly sharp teeth, must have been awesome to behold. No one, of course, had ever seen anything like it, except perhaps in their nightmares.

Steno had dissected sharks before, though never of this type or one nearly so large. He first examined the soft tissues of the head, which were already beginning to rot. With some effort, he cut through the tough skin and methodically peeled away layers of muscle and tendon. The translucent cartilage of the skeleton, delicately laced with purple blood vessels, gleamed like amber. Opening up the cranium, he was amazed by the smallness of the brain: barely three ounces for a body mass of well over a ton. How could such a tiny bundle of nerves direct the movements of such a huge creature? None of the glib theories of brain function he had denounced in Paris could explain this.

Then he turned the head over and gazed into the shark's mouth. Sharks such as these were known to be man-eaters, and Steno noted that this one's jaws could "let a whole man pass through them without difficulty." According to a story, which Steno did not doubt, one shark caught off the coast of France held in its belly a man clad in armor.

What particularly interested Steno, however, were the teeth. Steno recorded no dimensions, but those of a shark the size of Steno's would likely have blades up to three inches high. Many were missing, having been cut out on the beach during the excitement of

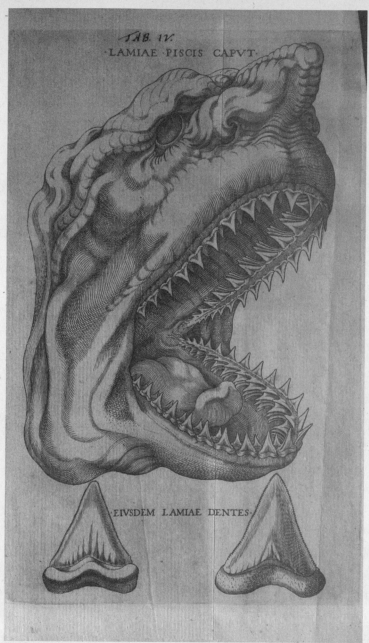

TAB. IV.

·LAMIAE·PISCIS·CAPVT·

·EIVSDEM·LAMIAE·DENTES·

Illustration of a shark's head and teeth from an unpublished catalog of the Vatican natural history collections. Although the curator of the collections, Michele Mercati, denied that tongue stones were shark's teeth, Steno later used this same illustration to affirm their identity. Nicolaus Steno, *Canis carchariae dissectum caput* (Florence, 1667). Courtesy of History of Science Collections, University of Oklahoma Libraries.

its capture, but hundreds remained. Each jaw held thirteen rows of teeth. Steno noted that the inner rows were soft and half-buried in the gums. He did not see what purpose they served. As for the outer rows, there was no question: the sharp, serrated blades were admirably designed for grasping the shark's prey and slicing it to pieces.

The teeth also gave Steno the chance to judge for himself a question that had been raised in the previous century by Guillaume Rondelet, a Montpellier physician and naturalist given to lingering in Mediterranean fish markets and buying from fishermen whatever oddities turned up in their nets. Rondelet noticed that the teeth of large sharks closely resembled certain unusually shaped stones called glossopetrae, or tongue stones. Could it be that glossopetrae were, in fact, the petrified teeth of sharks?

Rondelet's improbable suggestion must have raised more than a few eyebrows. It did not, as he claimed, necessarily contradict Scripture. But as a physician, especially one who grew up in a family of pharmacists, he surely knew it threatened other orthodoxies.

Like many other distinctive stones and fossils, tongue stones were prized for their curative, even magical, powers. Since ancient times they had been worn as charms to ward off everything from sorcerer's spells to infants' teething pain. Matchmakers claimed they were "very necessary" for "those that court faire women"; others credited them with the power to restore speech to the tongue-tied in general. Elixirs made from powdered glossopetrae were peddled as remedies for assorted "plagues, ill-disposed fevers, burns, poxes, pustules," as well as labor pains, epilepsy, and bad breath. The best and most famous glossopetrae came from the island of Malta, which exported vast numbers each year. Their special power was as an antidote to poison.

Because tongue stones were often found scattered on the barren ground surface, many people assumed they fell from the sky. Nobody had ever seen one fall, of course, which is probably why the Roman naturalist Pliny suggested it happened only on dark, moonless nights. Glossopetrae were most easily collected after heavy rainstorms, which led to the alternative theory that they were jagged shards of lightning bolts.

On Malta, the people knew that the glossopetrae did not come from the sky because they dug them out of the earth to sell. They had a more satisfying explanation, which they gleaned from a story in the Bible.

Once, while on his way to Rome, the Apostle Paul became shipwrecked on Malta. Cold and wet, he and his companions built a fire. Paul was stoking the fire when suddenly a deadly viper sprang from the firewood and sank its fangs into his arm. To everyone's astonishment, however, Paul calmly shook the snake from his arm with no ill effects. The Maltese believed that afterward he cursed the viper, depriving it and all its fellows on Malta of their venom. Ever after, nature commemorated Paul's miracle by producing stones, glossopetrae, in the shape of viper's teeth. The story thus neatly accounted for not only the stones, but their powers against poison.

Glossopetrae did not actually look much like snake fangs, nor did they look much like woodpecker's tongues, another hypothesis. But before Rondelet examined his French specimens, there were no better alternatives. Probably few people had seen the inside of a shark's mouth and lived to tell the tale.

Rondelet's shark-tooth theory did not attract much of a following after he published it in 1554, buried in a huge compendium of his other findings on Mediterranean fish. His comparisons between shark's teeth and glossopetrae were detailed and undeniably accurate, but he had neglected to explain how the teeth had turned to stone and what they were doing on land.

It was probably the glossopetrae's much-vaunted curative powers, not the question of their origin, that initially led an epileptic lawyer named Fabio Colonna to reexamine the question fifty years later. Forced to quit his law practice in Naples because of his illness, and not trusting doctors, Colonna launched his own experimental program of testing supposed cures, including glossopetrae. Casting a broad net, as it were, he read Rondelet's book on fish, with its account of shark's teeth.

Colonna's dogged studies of medicinal plants and stones had already established him as a first-rate naturalist and an independent thinker (Galileo was among his admirers). Rondelet had only considered overall shape. Colonna cracked open his glossopetrae to see their interiors, burned them to examine their ashes, and then did the same with teeth and with stones. In 1616 his "Dissertation on Tongue Stones" was published in Rome. It did not mince words: "nobody is so stupid that he will not affirm at once at the first insight that the teeth are of the nature of bones, not stones."

He was mistaken. Stupid or not, many people were unwilling to affirm any such thing. Glossopetrae were found on land, where it made no sense to find shark's teeth. Moreover, the shark-tooth theory could not account for glossopetrae's famous medicinal properties, which were a source of great civic pride to the Maltese, and no small comfort to glossopetrae users throughout Europe. In the end, Colonna was no more persuasive than Rondelet.

Exactly fifty years after Colonna went to print, Steno looked at the teeth of the huge shark before him. He was well familiar with Rondelet and Colonna's ideas; it's likely he had read them or at least heard about them while a student in Copenhagen. He had seen many glossopetrae in the fossil collection of his professor Bartholin, who had personally collected glossopetrae during a visit to Malta. Bartholin had, in fact, made his own study of glossopetrae.

Now Steno could draw his own conclusions. Bartholin never had the opportunity to compare shark's teeth and his Maltese tongue stones side by side, but Steno did just that, and it did not take much to see that their forms were as alike "as one egg resembles another."

It was also clear to Steno that the tongue-stone question was really only a special case of the general problem of fossil seashells and other "marine bodies" dug from the earth in places far from the sea. Shells and other marine fossils were often found side-by-side with tongue stones in rocks. The answer to the question lay in understanding how all of these bodies, not just tongue stones, came to be found there. And that meant studying not only the fossils themselves, but the places where they occurred and the materials in which they were embedded.

Since ancient times, people had found seashells in the Egyptian desert and in the mountains and hills of Greece and Italy. They had always been a puzzle. In recent years, as explorers and travelers had discovered shells in ever-higher mountains and greater distances from the sea, the problem had become acute.

"Nothing is so high, nothing is so far from the sea that we cannot find [shells] of those creatures that only live in sea water," marveled the Flemish traveler Jan Van Gorp in 1569. Flouting the conventional dread most people of his time felt toward mountains, Van Gorp bravely ventured up rocky summits accessible only by "goat tracks or by using ropes" in his search for petrified seashells.

"While in the Tridentine Alps and adjacent regions, I had difficulty in finding a way to the highest point of the mountains" he recounted breathlessly. "With each foot supported by six crampons and my hands helping me, with the assistance of an iron-shod pole, I myself found shells, and I learned from those who hunt capricorns, or ibexes, and deer, that these shells were often observed."

In South America, the Spanish mining official Albaro Alonzo Barba was similarly astonished to find seashells in rocks of the Bolivian Andes 13,000 feet above sea level. "I have some of these stones by me," he wrote, "in which you may see Cockles of all sorts, great, middle-sized, and small ones; some of them lying upwards, and some downwards, with the smallest Lineaments of those shells drawn in great Perfection."

It was obvious to him that "nothing but the Author of Nature itself could possibly have produc'd such a piece of Workmanship." Yet, like Van Gorp, he refused to believe that the shells could possibly be the genuine remains of marine animals: "This Place is in the Heart of the Country, and the most double mountainous Land therein, where it were Madness to imagine that ever the Sea had prevailed, and left Cockles only in this one part of it."

According to the thinking of the time, there were many more plausible explanations available for the growth of seashells in situ within rocks than for the flip-flopping of land and sea. The cosmos was a web of astral influences, sympathies, and occult powers. The earth was alive, tingling with "plastic forces" and "generative principles," dripping with "lapidifying juices" and "wet exhalations." Stones of every imaginable shape grew like plants; they fell with the rain, too.

In South America looking for gold, Barba was guided by the alchemical belief that each of the seven known metals was engendered in the earth by one of the seven celestial bodies. Gold, the brightest

and most noble of the metals, naturally was linked to the brightest
and most noble figure in the sky: the sun. Prospects were naturally
best where its rays were strongest, so Spain concentrated its coloniz-
ing efforts in the hot, sunny tropics, leaving the dank northern climes
for its rivals France and England. The riches they shipped home to
Spain seemed to be a tangible confirmation of this hypothesis. If ir-
radiations from stars and planets could penetrate the solid earth to
produce metals and ores, why not seashells?

Besides, an incursion of the sea could not explain all the *other*
strange things Barba found in Andean rocks: "the perfect resemblance
of Toads and Butterflies, and strange Figures, which tho' I have heard
from credible Witnesses, yet I forbear to mention, and not to overbur-
den the belief of the reader."

According to the old hypothesis of spontaneous generation,
seashells could spring up on dry land as readily as in the sea. "Tes-
taceans" (shellfish) and all other "non-copulatory" creatures, said
Aristotle, always reproduced this way. It was their nature to grow
spontaneously wherever conditions were ripe. Why shouldn't clams
and oysters sprout in salty desert soils and limey mountain rocks after,
perhaps, a good soaking by a rainstorm.

The Renaissance revival of Platonism and the mystical Hermetic
philosophies provided another ready explanation: the organic shapes
within rocks were produced in the same way as those of living, growing creatures. The characteristic features of an animal or plant species
came not from DNA or a genetic program—these ideas did not exist
in the sixteenth and seventeenth centuries—but were beamed into
them from the World Soul and the realm of Eternal Forms. The
Forms could just as easily mold the contours of a nascent stone as
they could an egg or seed. Of all the earthly realm, mountaintops

were the closest to the celestial sphere where the Forms supposedly resided. Why shouldn't they sometimes intercept "seashell" emanations headed for the sea bottom?

For many people, the best explanation was no explanation. Strange images in rock were part of a delicious riddle to be savored, but not necessarily solved. Nature had its playful, creative side; why shouldn't it adorn its mountains with whatever variety of shapes it wanted? Anyone who tried to analyze nature's jokes too closely just didn't get it.

Amazingly, human artifacts such as old pottery dug from the ground were also attributed to nature's mysterious fecundity. There was a certain logic behind this view. After lying in moist soil for centuries, clay pots and urns were often soft and fragile when first exhumed—only after drying and hardening in the sun did they become useable. Pottery hunters looked for them in the spring, "when they reveal their position by forming mounds as though the earth were pregnant." A reasonable person might very well look askance at the idea that the vessels were leftovers from some ancient people (of whom no historical records existed) and prefer to believe that they were fresh productions of the earth.

Collectors of curiosities displayed in their cabinets figured stones engraved with natural images of heavenly bodies, geometric figures, dragons, and castles, not to mention religious icons such as Christ, the crucifix, and the Virgin Mary. Next to such miracles—caused by chance patterns of coloration on rock surfaces, perhaps discreetly enhanced by chisel or paint brush—the natural sculpting of seashells in rocks was hardly surprising.

The problem, then, that faced Steno was not that there was no explanation for seashells and tongue stones inside rocks. There were *too many* explanations.

6

SECRET KNOTS

The world is bound in secret knots.

—ATHANASIUS KIRCHER,
MAGNETIC KINGDOM OF NATURE (1631)

The sixteenth-century enthusiasm for natural wonders had produced one great carnival of speculation over the origin of "figured stones." But the delight the Renaissance mind took in all forms of paradox and illusion was slowly being replaced by a more sober rationalism. Mysticism was out, mechanics were in. What was at stake was the lawfulness of nature. "The works of God are not like the Tricks of Juglers or the Pagents that entertain Princes," said Boyle the English experimentalist.

Immaterial forces, "sports of nature," and spontaneous generation were prime subjects of the new scientific scrutiny.

The reality of spontaneous generation had been thrown into serious doubt by Redi, but, as his critics were quick to point out, just because he had failed to observe it in his experiments didn't prove that it couldn't happen. Perhaps Redi had done something wrong, the conditions hadn't been right, the meat hadn't rotted properly.

Even Redi admitted that he couldn't rule out spontaneous generation in other kinds of situations, such as the growth of parasites inside *living* plants and animals. Internal parasites had always been a biological conundrum from a biblical perspective: no one could believe that there had been tapeworms in Eden, though one writer did suggest that before the Fall, they had been benign, and only "gently licked" the intestines of Adam and Eve. Most people, however, assumed that parasites were part of the curse of mankind, coming after the original Creation, and proliferated spontaneously within all the corrupted flesh of the world. How else could they appear deep in the tissue of other living things? Only now were investigators like Swammerdam and Marcello Malpighi discovering through their microscopes that even these tiny "imperfect" creatures had reproductive organs and produced their own minute eggs.

For fossil seashells to grow inside rocks by whatever cause implied a randomness and purposelessness in nature contrary to Steno's scientific and religious beliefs. As an anatomist Steno believed that every biological structure was precisely designed for a particular function— his studies of the lowly glands discredited the Aristotelean maxim that some organs were more perfect than others—and as a devout Christian he believed that God created nothing without a purpose. From a mechanical standpoint, things similar in structure must also be similar in function. If they were produced by a natural process, they must also be similar in origin.

The different theories that purported to explain fossils by their growth inside rocks were hopelessly vague on just this point—how it happened. They were based more on religious and philosophical premises than on the structure of either the fossils or the places where they supposedly grew. Steno's dexterity with a scalpel would be of no help here, but his habit of precise observation would.

In the report to Ferdinando, which he wrote immediately after the shark dissection, Steno added a brief "digression" on the origin of tongue stones and other fossils. There had been little time for him to investigate the matter thoroughly, but he offered a few observations and "conjectures" anyway.

One immediate problem he saw with any theory that claimed fossils grew inside rocks was that in none of the places where he had found them did they seem to be growing in the ground there today. A shell growing in solid rock would necessarily crack the rock, but he never saw cracks around fossil shells. And the shells in softer ground not only showed no signs of growing, but appeared instead to be slowly disintegrating. Water seeping through the soil had often leached and dissolved shells to the point that they could be easily crushed by the fingers. On the slopes where people found fresh batches of shells each spring, claiming that this proved the shells were newly generated by the earth, that was easily explained by rain and erosion winnowing away soft clay and soil, leaving the fossils as a residue.

What was more, things *known* to grow in the soil, such as tree roots, inevitably became "twisted and compressed in countless ways in harder ground, so that they assume shapes different from those in softer ground." But glossopetrae and fossil shells were always the same, "whether dug up in softer ground, hewn from rocks, or taken out of animals."

Fossils were also different from minerals. Mineral crystals were relatively simple geometric shapes, basically just combinations of cubes, pyramids, and other polyhedrons. Yet it was extremely rare to find a crystal without some kind of defect, such as a truncated corner or a warped side. This was true for not only crystals found in the earth, but also those of salt or ice grown by experiment. But what about tongue stones and other marine bodies found in rocks, whose shapes were vastly more complex and irregular than inorganic crystals? "How much greater and more numerous should be the defects observed in bodies possessing a much more compound shape, and in those that are copies of animal parts." Yet fossil shells were always perfect replicas of their biological counterparts "in the course of the ridges, in the texture of the lamellae, in the curvature and windings of the cavities, and in the joints and hinges of the bivalves."

Even their flaws perfectly matched the sorts of chips, breaks, borings, and scratches one could see on real animal remains: "If certain mussel shells are found broken across the middle, the edge of the fragment itself provides evidence that another part was once attached to it; indeed, this is often found close to the first." What force of nature would cause a piece of shell to grow in a jagged shape that would fit, jigsaw-puzzle style, into another shell buried a few inches away? And when shells or teeth were found in clusters, they were not randomly growing out of one another in the manner of crystals, but in the same kind of arrangement that would be seen on the seafloor or a beach. "If several tongue stones of various size, not all of them complete, are observed sometimes to stick together, as if in the same matrix, the same is noted in the jaw of a living animal where neither are all the teeth of the same size nor are the teeth arranged in the inner rows completely hardened."

The exact resemblance between fossils and their living biological counterparts, and Steno's cogent arguments against their growth inside the earth, would seem to be definitive. Steno must have been convinced. But in his report to Ferdinando, he is strangely noncommittal. "It is easy to show that the shape of those bodies is no obstacle to our considering them to be parts of animals," he writes. It seems to him that those who take the position that glossopetrae are shark's teeth are "not far from the truth." But he is also careful to make the disclaimer:

> *While I show that my opinion has the semblance of truth, I do not maintain that holders of contrary views are wrong. The same phenomenon can be explained in many ways; indeed Nature in her operations achieves the same end in various ways. Thus it would be imprudent to recognize only one method out of them all as true and condemn all the rest as erroneous. Many and great are the men who believe that the said bodies have been produced without the action of animals.*

Part of this may have been laudable scientific caution, and part of it courtly decorum. This was a new topic of research for him and he was a newcomer in a country where fossils were plentiful and learned debate about them had been going on for centuries.

It may also reflect his uncertain status at the Medici court. The previous summer he had been warmly welcomed by Ferdinando as an honored guest, immediately involving himself in the Cimento's activities, helping out with experiments, and adding his ideas to their philosophical discussions. A permanent appointment to the court would be a plum for any aspiring scientist. If Steno was hoping to impress Ferdinando, he could not afford to embarrass himself by drawing hasty conclusions before he had thoroughly investigated the question. He was, after all, going against the opinions of the "many and great" who supported the in situ theories.

First among these was the near-legendary Jesuit polymath and collector of natural curiosities Athanasius Kircher. Just the year before, Kircher had published his authoritative treatise on the earth and all its productions: *Mundus subterraneus*, "The Subterranean World."

Hardly remembered today, Kircher was a giant among seventeenth-century scholars. Straddling the divide between the expansive scholarship of the Renaissance and the focused data-collecting of the emerging scientific age, he was one of the last thinkers who could rightfully claim all knowledge as his domain.

Kircher's museum in Rome was easily the most famous natural history collection in Europe, not only for its vast holdings of fossils and other rarities (including such dubious relics as the rib and tail of a mythological Siren), but for his exuberant sense of showmanship. Kircher's museum was no mere repository of curious objects, but, as he put in an aptly titled book, *The Great Art of Light and Shadow*, a "sensory Theater of the World."

Visitors to the museum were greeted by Kircher's disembodied voice, which came through a hidden metal tube as he spoke from the privacy of his bedroom. Once inside, they were free to peruse the floor-to-ceiling displays of natural objects—or watch a magic-lantern show of projected images, or marvel at Kircher's many mechanical inventions (including perpetual-motion machines), or pose questions to an animated statue, the "Delphic Oracle." The Oz-like Kircher supplied the statue's responses through another speaking tube, while moving its mouth and rolling its eyes at visitors' queries.

Kircher the man was, if anything, more impressive than his museum. A German-born Catholic, he had traveled Europe in his youth during the worst of the Thirty Years' War, surviving trampling by horses, capture by Protestant soldiers, near-fatal disease, several shipwrecks, and, on at least one occasion, being washed through a mill's waterwheel.

In 1634, still miraculously in one piece, he turned up in Rome. Pope Urban VIII, who had condemned Galileo just the year before, hired Kircher on the spot as a translator of Egyptian texts. Before long, he was Professor of Mathematics, Physics, and Oriental languages at the Jesuit College of Rome. Finding himself at the center of a worldwide network of Jesuits from China to Mexico City, he became a one-man information explosion, writing tome after tome on every imaginable subject, sacred and profane. His bedroom collection of curios grew into a museum, which soon bulged with antiquities and natural objects sent by missionaries in Asia, Africa, and the New World. He mastered twelve languages; all Rome was dazzled. "What universality! What profundity of knowledge!" exulted one admirer.

He became, in effect, the church's response to the Galileo debacle. Dutifully affirming the traditional earth-centered cosmos, the good priest Kircher devoted his prodigious energy to restoring Rome's reputation as a center of intellectual activity. His passions were magnetism and Egyptian hieroglyphics—both of which he claimed had profound mystical significance—but he also wrote about optics, astronomy, music theory, mathematics, theology, and everything else that captured his fancy. Always ready to put science into service for the faith, he helped keep the chapels full on Sunday by taking his magic lantern out into the streets at night and flashing images of the devil through the windows of homes to terrify the inhabitants. He invented a computer and a "cat piano." (Without going into details, the latter involved sharp needles and cats with differently pitched meows.) A rumor spread that he had also discovered a way to fly, but refrained from doing so at the church's request.

Despite his vaunted erudition, Kircher retained a childlike manner that others found remarkable. "He is likewise one of the most naked and good men that I have seen, and is very easy to communicate whatever he knows, doing it, as it were, by a maxim he has," an

English visitor wrote home. "On the other side he is reported very credulous, apt to put in print any strange, if plausible story that is brought unto him. He has often made me smile."

Kircher's interest in the inner workings of the earth was spurred in 1638 by a hair-raising escape from an earthquake while he was traveling in southern Italy. It was a terrifying spectacle: The ground heaved so violently that it knocked him off his feet, towns and castles collapsed around him, and volcanoes spewed smoke and flame.

But along with the fear came fascination. Afterward, the landscape still in ruins and the volcanoes still smoldering, Kircher hiked up Mount Vesuvius and had himself lowered by a rope into its crater. "It was terrible to behold," he reported. "The whole area was lit up by fires, and the glowing sulphur and bitumen produced an intolerable vapor. It was just like hell, only lacking demons to complete the picture."

Kircher was overwhelmed. "O God," he exclaimed, ". . . how incomprehensible are your ways!" Twenty-six years later, however, he decided he had the basic ideas worked out—and published *Mundus subterraneus.*

Printed in two massive, copiously illustrated folio volumes, *Mundus subterraneus* was an encyclopedic amalgam of geologic fact and fable—offering not only Kircher's views on volcanoes, mountains, caverns, and rivers, but a recipe for gunpowder, advice for treating snakebite, and computations on how long it would take a swallow to fly around the world.

What of figured stones and other images in rock? Kircher saw not only seashells and other creatures, but an infinite variety of shapes, including, as he illustrated in his book, the complete Greek and Roman alphabets, etched by nature into stone.

Like earlier Renaissance thinkers, Kircher believed that the natural world was a matrix of signs and symbols freighted with mystical significance. The goal of his science and his tireless collecting of natural objects was a total, intuitive, universal knowledge of the cosmos. To his mind, nature, like Scripture, was sometimes best understood metaphorically as a clue to some hidden, deeper meaning.

In designing the world, God attended not only to the physical needs of human beings, but their spiritual needs as well. Just as Kircher with his magic lantern had projected images into people's houses, God in his wisdom put images in the earth to guide the faithful.

Why was the lion put on earth? To teach us courage. Why the ant? To teach us industry.

Why were seashells put in rocks on top of mountains? Well, that was less clear. But the ultimate goal of all Kircher's science and scholarship was an intuitive awareness of the invisible cords that tie all things together—heaven to earth, land to sea, and so on. Kircher disdained the narrow, mechanical explanations of the new philosophers.

In explaining just *how* fossil shells got into rocks, the openminded Kircher refused to restrict himself to any specific hypothesis. He freely granted that petrified shells in low-lying deposits might be genuine shells buried by minor inundations of the land. He went farther than that, even. Petrified creatures abounded in his world, as did petrified humans. In *Mundus*, he wrote of a "whole village in Africa turned into stone, with all the people thereof."

But if some marine fossils on land were left there by the sea, those in the high mountains were the work of other forces. These included astral emanations, spontaneous generation, and, his favorite, the mysterious force of magnetism. Fossils took their characteristic shapes from what Kircher called a "plastic spirit." Permeating the earth, it was this plastic spirit that gave form to all things animal, vegetable, and mineral. The plastic spirit was active to this day, generating flies in dung (regardless of what Redi believed), and molding shells and figured stones all through the earth's crust.

Though much of Kircher's theory about the growth of fossils in rocks was warmed over from previous medieval and Renaissance writers, he did not rest on the authority of any text, excepting of course the Bible, to back up his assertions. He knew that although Galileo had lost the battle with the church over the solar system, he had won the larger war in science about the primacy of the experimental method. Kircher was no simple convert to this new method, he was (in the words of a pupil) "the prodigious miracle of our age who has excited the admiration of the whole world by the innumerable experiments on which he has based his universal sciences."

Admittedly, Kircher had not yet created even a single fossil clam in the laboratory. But in *Mundus subterraneous,* he claimed to have experimental proof that the plastic spirit was a real force of nature. Most were spontaneous-generation experiments, which he could get to work even if Redi couldn't. The growth of seashells inside rocks, then, had a firm scientific basis, newly endorsed by the most illustrious of authorities.

TESTING THE WATERS

How well then everything fits together! How unanimously they come together in agreement. We find the position of the soil suited to its having been able to hold waters; we know that both powdered soil and the elements of the said soil could have been mixed with the waters; we do not ignore the ways in which they could have both entered and separated from those waters, nay rather we pay close attention to the variety of layers in the soil itself. Why then is it impossible for this soil to have been a sediment from water?
—NICOLAUS STENO, *A SHARK-HEAD DISSECTED*

I n the world of Athanasius Kircher, seashells inside of mountain rocks could not possibly be real seashells because mountains were created before mollusks. The difference was just a few days, but that was enough.

The origin of mountains had always been a topic of speculation. Genesis does not mention them at all until the story of the Flood. To many readers this meant that, like tapeworms, they weren't part of the original plan. Only after Adam's fall did they well up from the ground, "even as Warts, Tumors, Wenns, and Excresencies are engendered in the superficies of men's bodies." This was just how many people saw them. *"Vast . . . horrid, hideous, ghastly ruins,"* was a typical seventeenth–century tourist's reaction to Alpine scenery.

Kircher took a more sanguine view. To him, the world was a marvelously integrated system: all its parts had to have been created together, at once, fully functional and in their present conditions. That included mountains. He believed that they formed the earth's skeleton, keeping it from collapsing into a shapeless blob of sand and mud. In *Mundus subterraneus,* he offered as proof the dubious claim that the world's mountain chains were arranged in a grid, some tending north-south, the others east-west.

Kircher allowed that the oceans occasionally overflowed, and that heaps of mud and sand left behind could form hills. But the high rocky mountains were essential to the world's integrity. Obviously, then, these mountains, and the rocks holding them up, had to have been created "by the divine architect at once at the beginning of things." That was on the third day, when God scooped out the ocean basins and piled up the dry land. Seashells and other marine life weren't made until the fifth day.

Steno was no less reverent toward the Bible than Kircher, but when he looked at rocks that held fossils, what he saw seemed to imply a different sequence of events. The undistorted shapes of the fossil shells implied that when they became entombed, the rock was not yet solid. That is, it was not yet rock. The shells did not grow inside the rock, the rock solidified around the shells.

To visualize how this could happen, Steno thought back to experiments he had seen in Ole Borch's laboratory. It was easy to dissolve various powders in water and other fluids. Then later, after the water had cooled or partially evaporated, solids would precipitate and settle to the bottom particle by particle to make a soft ooze. After some time, the ooze would harden into a solid mass.

A similar process could happen in nature. A rock containing fossils could have originally been a soft sediment laid down by water. A fossil clam might have lived, burrowed into this sediment, while a fossil mussel might have lived on its surface. But either way, the rock was sediment first. Only later had it hardened into rock.

Water was the key ingredient. Mixed in with the sediment, it not only accounted for the deposit's original softness, allowing the shells to grow easily without distortion, but it also provided a supply of sediment. If the fossil-bearing hills were to be seen as big heaps of sand and mud, the sand and mud had to be brought there somehow.

A miraculous flood would certainly do the job, but as Steno pointed out, there were other, more mundane, possibilities. Just the normal processes of erosion that anyone could observe inevitably washes soil toward the sea "That clay and sand are mixed with strongly agitated water is so obvious from the headlong course of torrents . . ." he said, "and from the agitation of waters by the wind, that no further explanation is needed." And of this "muddy water, either from the ocean or from torrents, it is certain that the bodies which make the water muddy ought to sink to the bottom when the violent motion ceases. Nor do we need to seek diligently for examples of this type, since both the beds of rivers and their estuaries give sure proof of it."

Indeed, a closer look at the deposits themselves gave another clue that they had been deposited by water. When agitated, sediment-laden water quiets down, the particles of sediment sink to the bottom in order of their weight—larger and heavier particles drop out first, fine silt and clay remain in suspension much longer. Steno had seen the same arrangement of sediment grains in the deposits where fossils were found. This seemed to prove that they had been laid down by water.

Often, however, the rocks in which he had found fossils did not bear such an obvious resemblance to sediments. Many fossil deposits are limestones, which can have a smooth or crystalline texture, without any recognizable sedimentary grains. Limestones are typically laid down in warm, tropical seas—seas that at the time had barely been explored by Europeans. Steno certainly had no firsthand experience with the kinds of places where these kinds of sediments were laid down.

Yet Steno recognized that the familiar mud and sand swept into the sea by rivers might not be the only kind of sediment that could entomb fossils. Sediment could also come from minerals dissolved in water as well as grains carried by turbulence. Rivers are fed by springs, which exude from the earth. Water wending its way through subterranean passages would naturally dissolve minerals, which could later precipitate as sediment. One could even observe this process happening on the dripstone formations of caves.

Therefore, Steno supposed, the different rocks of the earth were made from different sediments, some washed off of the land and some precipitated from dissolved minerals, which had accumulated in great thicknesses on the sea bottom. Fossils were former creatures who had lived there, their remains buried by the slow buildup of sediment.

If rocks were formed out of water-laid sediments, then the observation that the types of fossils found in rocks were normally aquatic creatures, was not an anomaly at all, it was to be expected. The types of creatures most likely to become fossils are obviously those that live where rocks are made. And the fact that fossils were usually the durable parts of animals—shells, bones, and teeth—also supported his idea that the sediments accumulated gradually. Soft, perishable tissues would disintegrate too quickly to become buried and fossilized.

Steno's interpretation agreed with the fossils and the rocks. It was harder to reconcile with the present-day landscape, but Steno reasoned it could happen in one of two ways: either the water rose to cover the land, or the land sank beneath the water. Both seemed possible to Steno. Scripture affirmed that water could cover the land. In fact, it had happened at least twice: at Creation and during the flood of Noah. And in the writings of the ancients there were many accounts of mountains raised and lowered by spectacular earthquakes. The bedrock itself, often cut by fractures and fissures, testified to past upheavals.

Therefore, he concluded cautiously, "if anyone should believe that portions of soil in places from which the said bodies have been dug have changed their situation at some time, he cannot be held to think anything that is contrary to reason or experience."

The island of Malta, motherlode of tongue stones, gave a prime example. "If we believe the accounts, new islands have sprung up from the depths of the sea, and who knows Malta's origins?" he wrote. "Perhaps at one time when this land lay under the sea, a place where sharks lurked, their teeth were buried frequently in the muddy sea-bed, but now they are found in the middle of the island owing to a change in the position of the sea-bed."

The perfectly designed, immutable world might not be so immutable after all.

In the spring of 1667, Steno published his *Canis carchariae dissectum caput,* "A Shark-Head Dissected," with its cautiously worded arguments on fossils and sediments as an addendum to the muscle treatise he had completed earlier that fall, still waiting at the printers. To an otherwise dry and long-winded anatomical discourse, he thought it might add some "spice."

Spice or no, Ferdinando must have been very pleased with what he read because soon after it came off the presses, he granted Steno lodgings in the palace and a modest pension. Steno would henceforth be treated as a full-fledged member of the Cimento. He was free to

NICOLAVS STENONIVS

Portrait of Steno, in 1667 or 1668. The original was painted while he lived at the Palazzo Vecchio, probably by the Medici court artist Justus Sustermans. Courtesy of the Institute of Medical Anatomy, University of Copenhagen.

pursue his curiosity wherever it led him, and Ferdinando would gladly pay all expenses.

Steno now began the work that ultimately led him to be called the "founder" of geology. Considering the disparate intellectual threads he had to pull together, the conceptual barriers he had to overcome, and the distractions that intruded into his life, he worked with amazing speed. The time from the capture of the shark to the composition of his masterpiece spanned less than two years. But Steno complained about the slowness of his progress.

He did not immediately give up his anatomical research. In fact, during this same period he made one of his most important discoveries. He found that females of live-bearing species produced eggs, just as egg-laying ones did. Anatomists had previously assumed that the ovaries seen in female cadavers were simply "degenerate" testes. Steno's finding, confirmed a few years later by his friends Jan Swammerdam and Reiner de Graaf (who fought bitterly with each other over credit for it), overturned a long-standing myth of human reproduction: that the female was simply a passive receptacle for the male seed.

But as Steno became more and more engrossed in geological studies, taking long "mussel journeys" into the mountains to collect fossils, he let other projects slide. The follow-up study of the body's muscles was completely abandoned, and reports on earlier dissections remained unwritten.

As a student he had agonized about his shifting focus, but now he had more-or-less come to terms with it. He told Ferdinando, who for two years patiently allowed him to follow his wandering curiosity, that he wondered if behind it was a "higher cause."

Whatever was behind it, the new direction his research was taking was proving to be far more complex than anything he had tried in the past. For all its flaws, anatomy was nonetheless an established science, with clearly defined boundaries and known methods of research. There were elder colleagues and a vast body of literature he could turn to with questions. Most important, in anatomy the object of interest—the human body—was, for all its mysteries, right there on the table in front of the investigator.

But his fossil studies raised new and unfamiliar kinds of questions. Many were tied to things such as ancient seas, that no longer existed, and happenings deep in the earth, which could not be observed.

Steno's investigations took him back and forth across Tuscany. He traveled with Ferdinando's court to hunting grounds in the mountains and to seaside villas near Pisa. Florence's innumerable stone quarries gave Steno glimpses of the Tuscan landscape's bare flesh, as it were, beneath its skin of soil and vegetation. Mines in the metalliferous hills, worked since Etruscan times, were opened to Steno at Ferdinando's behest. Even the stones of buildings, the slabs paving the streets and piazzas, gave clues.

From what he saw on these excursions, he was all the more convinced that his original conclusions were correct, that the "soils from which shells and other marine deposits are dug" were indeed "sediments from a turbid sea," and moreover that from this it was possible "to calculate in each place how many times the sea had been turbid there."

Still, he was struggling to put together a coherent picture of how all the different kinds of "bodies" he observed could come to be enclosed in the rocks. "As I investigated more closely both each place and each body," he said, "these gave rise to a succession of doubts, indissolubly connected, which assailed me day by day."

It did not help that at the same time he was being assailed by other events.

The Cimento was breaking up. For several years, relations within the group had been deteriorating. Much of the problem could be traced to the brilliant but quarrelsome Giovanni Alfonso Borelli. Borelli had no patience with his intellectual inferiors, and was suspicious of his few equals. Under Leopoldo's protection, Borelli had just published a treatise on the forbidden Galilean astronomy that in some ways anticipated Newton. Some years before, he and Viviani had been the first to measure the speed of sound; now they were barely on speaking terms.

Though Borelli shared Steno's interest in the mechanics of muscle contraction, he was plainly unhappy when Steno arrived in Florence. Steno's offers to collaborate were rebuffed. Borelli feared the younger scientist would "skim the cream" from the subject before his own work was finished.

As it happened, Borelli's research on muscles was not published until after his death. Although it does not mention Steno by name, his posthumous book vehemently attacked Steno's idea that muscular action comes from the contraction of muscle fibers, not the ballooning of the entire muscle mass. Discredited in the eyes of contemporary scientists by Borelli's assault, Steno's theory, radical at the time but now known to be correct, had to be independently rediscovered by physiologists almost a century later. Not until the 1980s would scientists looking back on Steno's work recognize the prescience of his muscle studies.

In Borelli's battles with other members of the Cimento, a major point of contention had been the academy's sole publication, the *Saggi di naturali esperienze*. In Leopoldo's idealistic view, the search for scientific truth should be a disinterested and communal enterprise. The upshot was that the *Saggi* would be published anonymously, and would only include facts and conclusions on which all members of the group agreed. For several years Lorenzo Magalotti, who had been enlisted to draft the *Saggi*, had suffered through endless revisions, no two scientists agreeing precisely on what their experiments had proved, how to describe them, or even how to punctuate Magalotti's elegant Tuscan prose. Borelli bristled at having to share credit for ideas he saw as his, and, worse, having to compromise them to please the rest of the group.

Now that the *Saggi* was finally finished, Borelli announced he had had enough and was leaving the Cimento to retire to his native Sicily. Two other academicians followed suit and resigned.

Losing Borelli, who, for all his prickliness had been a driving force among the scientists, was a devastating blow to the Cimento, but not fatal. It had, after all, just acquired Steno. What killed the academy was the news that Prince Leopoldo himself would also soon be out of the picture as well. The church had selected him to become a cardinal.

The freethinking Leopoldo was a surprising choice for a cardinal as his penchant for reading books prohibited by the church was well known. At the Cimento, he and his academicians joked openly about "convict[ing] the Book of Genesis of errors in several places."

But politics was politics. For years, Ferdinando had been negotiating with Rome about a cardinalate for the Medici family. Relations between the Medicis and the Vatican had been uneasy since the Galileo affair, with the offer of a cardinalate, the newly elevated Pope Clement IX was hoping to patch things up. The original choice may have been Leopoldo's older brother Mattias, but he fell ill and died in October 1667. Leopoldo was thus the only male Medici available, and two months later, he was elected.

Leopoldo's duties as a cardinal did not promise to be so onerous that he couldn't keep up with scientific interests, or preclude the "contemplative life" he enjoyed so much. But it would preclude his overseeing a research institution. And without Leopoldo, the Cimento was dead.

Just a month before Leopoldo's election in Rome, Steno's own program of scientific research was also abruptly thrown into question. He had received the following letter from the king of Denmark:

> *Our favor as of Yore! Mayest thou know that We until further notice by singular grace and favor have most graciously advanced to thee 400 rixdollars annually, which pension shall begin and have respect to the time thou shalt be returned hither, for it is Our gracious will and command that thou shouldest with all dispatch set out on thy journey hither to the Realm of Denmark so that thou canst be home immediately or in the first coming spring. Whereafter thou, etc.*

The note was a little vague on what was expected of him on his return. Kings were not in the habit of explaining themselves, but it was clear that the sovereign had commanded his famous subject to return home immediately.

The job offer could have been a choice opportunity for Steno. Being attached to the royal court promised great prestige and an ample salary. In the past, Steno had attempted to curry favor with the Danish royal court: He had dedicated his first major work on muscles and glands to the king. Presumably his friends at home—especially Bartholin and Borch—had long been lobbying for the creation of some kind of scientific position for their illustrious countryman and former student.

But given his present scientific interests, it put Steno in an awkward position. Should he return to Denmark, his geological studies would have to go on the back burner, perhaps forever. He would have his hands full with anatomical work. And, besides, flat, sandy

Denmark hardly offered the same opportunities for geological investigations as did Italy with all its mountains, mines, and volcanic fumaroles.

There was also another, more serious problem. Steno had done something the Danish king might find unforgivable: That very day, he had forsworn the faith of his homeland and officially had become a Catholic.

CONVERSION

The heart has its reasons which reason knows nothing of.
—BLAISE PASCAL, *PENSÉES* (1670)

For Robert Boyle, it happened during a lightning storm so violent that he prayed for his life. Shortly thereafter, he took a vow of chastity and resolved to use his science to further Christianity. For the mathematician Blaise Pascal, it came in a "night of fire." He recorded his feelings on a scrap of parchment:

> Fire
> *God of Abraham, God of Isaac, God of Jacob, not of*
> *the philosophers and scholars.*
> *Certitude. Certitude. Feeling. Joy. Peace.*

Pascal wore the parchment stitched into his clothing every day for the rest of his life.

Steno's experience of religious conversion lacked the pyrotechnics of the other two scientists, but was no less soul-wrenching. He later told friends it was triggered by a chance event as he walked a

Florence street "at eventide" on All Souls' Day, November 2, 1667. A woman's voice called to him from a window: "Go not on the side you are about to go, sir, go on the other side." She was directing him to his friend's house across the street. But "that voice struck me," he said, "because I was just meditating on religion."

For many months he had agonized about crossing over to the Catholic church by talking to friends and reading. With characteristic punctiliousness, he compared the theological claims of Protestantism and Catholicism point for point by exhaustively searching original Greek and Hebrew manuscripts in the Medici library rather than relying on Latin translations of the Bible. For all his studies, though, it was not in the end a rational decision. That day on the street, he said, everything suddenly became clear and he cried out: "O Lord, Thou hast broken my chains asunder!"

Steno's "chains" were the religious doubts that had gnawed at him since Amsterdam. Like Pascal, he craved certitude, but certitude was the one thing that the new philosophy could not provide. "If a man will begin with certainties, he will end in doubts," said Francis Bacon, "but if he will be content to begin with doubts, he shall end in certainties." But for Steno, doubts had only led to more doubts.

One event had particularly shaken his sense of intellectual security. It happened while he was a student in Holland. Despite his conservative upbringing at home, Steno had gravitated toward some of the more freethinking elements of the Dutch intellectual scene. They were far more interesting than the older, more conservative academics still beholden to Aristotle. Among this crowd were some of the brightest lights of the era, notably Baruch Spinoza, the philosopher. Spinoza, living in the small town of Rijnsburg, occasionally came into Leiden to watch Steno dissect. The two would discuss the philosophy of René Descartes.

Steno had been introduced to Cartesian philosophy by Ole Borch, and since that time had read as much on Descartes and his works as he could lay hold of. He admired Descartes's "geometric" reasoning and his critical "method of doubt," that is, doubting every proposition until it has been positively proved. Descartes had promised that his methods would lead to certain knowledge, not only in philosophy, but science and religion as well. When Descartes's posthumously published book *On Man* became available, Steno, as a student of anatomy, immediately got hold of a copy.

Cartesian anatomy was a product not so much of observation as of reason. Descartes described the body's structures not as he saw them, but as they *must necessarily be*, given the constraints of his system of geometric reasoning and the mechanical view of anatomy.

The keystone of Cartesian anatomy was the theory of the heart. The dispute had always been whether the heart was a muscular pump, forcing blood into the blood vessels during its contractions, or if it was a furnace, generating the body's heat and causing the blood inside it to violently expand and rush into the vessels.

To Descartes, the heart had to be a furnace. It was impossible for it to be a pump. The heart lay deep inside the chest, isolated from all of the body's other muscles and moving parts. There was no mechanical way for motion to be transmitted to it, and for the heart to generate its own movement violated the fundamental tenets of the mechanical philosophy. The heart was therefore not a muscle, and so the function of its tissue was to generate heat, not movement. Descartes went so far as to say anyone who rejected his theory of the heart must reject his entire philosophy.

Soon after reading the book, Steno and a friend dissected an ox heart to see how it squared with Descartes's contentions. After boiling the heart to soften it and peeling away its outer membranes, Steno found that its walls were fibrous like a muscle. He saw also that the fibers were arranged in such a way that by shortening they would squeeze the heart precisely as it would normally contract to force blood into the arteries—just like a pump.

This was a shock, Steno said, because up to that point he had held Descartes to be "infallible." Wanting to be certain of his interpretation, he decided to compare the heart tissue against a normal muscle.

"Happening to have a dead rabbit at hand," he wrote, "I laid hold of its legs and separated its muscles." It was immediately clear, "in the first muscle and by the first cut," that there was no fundamental difference between heart and muscle tissues.

His faith in Cartesian science was utterly destroyed. If the "indisputable evidences" of Descartes and his followers were errors "which in an hour or so I can get a ten-year-old to demonstrate," said Steno, "what certainty can I then have about other subtleties of which they boast?"

Steno's disillusionment extended beyond science. The Cartesian philosophers claimed a rational proof of the existence of God. But "if they can be so mistaken about material things, which are accessible to the senses, what certainty do they give me that they are not also in error when they state their views about God and the soul?"

The philosophical crisis hit Steno at a vulnerable time in his life. Dutch urban society, with its overwhelming social diversity, had already delivered the first shock to the bedrock of his faith. Holland was a rare enclave of religious liberalism in a Europe that was otherwise bent on achieving uniformity. The conventional solution to religious conflict was to stamp out dissent. In England, Marvell the Puritan derided the freedoms given to Amsterdam's motley population: "Turk-Christian-Pagan-Jew, staple of sects, mint of schism." But the Dutch were unapologetic. Liberalism was less a matter of idealism for those in power than of practicality. Tolerance was good for business. Amsterdam merchants sold books banned elsewhere, and they were willing to do business with anyone. A typical Amsterdam trader, it was said, would sail through hell if he thought it would return a profit.

Growing up in uniformly Lutheran Denmark, where its truth was unquestioned, Steno had been overwhelmed by the religious diversity he had encountered in Holland. Religion was supposed to guide one's life, to lead one to charity, temperance, and love in this world and salvation in the next. People held different doctrines and worshiped in different ways. Regardless of the values that were preached, he saw little difference among the various faiths in the sanctity of the lives of their respective believers. In particular, the depraved Catholics he had expected to meet turned out to be no different from anyone else.

Yet everyone was convinced that their faith was the one true faith. They could not all be right. Did that mean they were all wrong? Was there such a thing as "the one true faith"?

Steno's discomfort was not unusual among educated Europeans in the wake of the Reformation. A century of doctrinal warfare between Protestants and Catholics had left many feeling caught in the crossfire. When Protestants claimed that religious truths lay completely and exclusively in the literal words of the Bible, Catholics pointed out that its text was fraught with internal contradictions, its true meaning often obscure and metaphorical.

Moreover, Scripture existed in three versions: the Greek, the Hebrew, and the Latin. Which "plain words" was an individual reader supposed to choose? But when Catholics claimed that the authority to interpret Scripture must rest in a central authority—the Church—Protestants were quick to point out that the Church legitimized its claim by appealing to Scripture. If the authority of Scripture was to be doubted, then so too must the authority of the Church be doubted.

For many people, this undermined the whole idea not just of faith, but knowledge of any kind. Not only were books mere lies and fables, but the evidence of the senses was unreliable; logic an empty game; mathematics an esoteric dream. There was no right, no wrong, no truth, no falsehood. To these people, the godless lawless Epicurean universe, with its underlying randomness and atomism, seemed to be borne out by the chaos they saw around them. There was no sense to be found in the universe, so there was no sense in looking for any.

The Scientific Revolution had seemed to offer an antidote to the confusion. It was as much a reaction against the radical skepticism prevailing in some quarters as it was against the dogmatic certainties of the Scholastics. When Descartes proposed his "method of doubt," it was because he believed, like Bacon, that it was the path to certainty. After erroneous ideas were discarded, he said, what remained would be certain.

The idea that certain knowledge was possible came to Descartes more as a religious revelation than as a product of the kind of deliberate reasoning he advocated. One cold night, as he lay in his famous "stove-heated room," a series of visions came to him. In the first he was caught up in a whirlwind. In the second, his room filled with a shower of sparks. In the final vision, a mysterious figure appeared and handed him several books, which he understood to represent knowledge. Somehow, Descartes extracted from these visions his mission in life: to create a universal system of knowledge based on mathematical reasoning. After all, it was only in geometry that knowledge was universally admitted to be certain.

Geometry was the solution to the crisis of faith and knowledge.

"God is a geometer," said Kepler. Reason would lead to intellectual proof of God—absolute *certainty* based on reason, not authority or simple faith.

Out of this urge to find God in the order in the universe, not from the authority of texts, came the idea of "natural religion." All religions grew out of a religious impulse implanted in the heart by God himself. One needed only to look into one's own self, and through the proper application of reasoning, one could derive everything one needed to know to attain salvation. No "revelation" was necessary. The Bible was surely inspired by God, but it was simply the "hard copy" of what an introspective person could find out on his own.

But for Steno, the evidence of an ox heart and rabbit leg had demolished all his hopes that there ever could be a reason-based religion. Like Pascal, he would come to believe that God could be found only by a blind leap into the infinite. What the sober-minded, analytical Dane wanted, as it turned out, was the kind of emotional intensity he finally found in Italy.

Soon after arriving in Italy, he had been overwhelmed by the spectacle of a Corpus Christi procession in Livorno. "Either this host is no more than a piece of bread, and they are fools who pay it such homage," he said, "or it really does contain the body of Jesus Christ, and why do I not also honor it?"

He was impressed by the zeal of Italian Catholics, particularly characters like Paolo Segneri, a Jesuit preacher equally famous for his skill as an orator and his capacity for self-inflicted penance. Segneri walked barefoot between speaking engagements, wandering three hundred miles each year to spread his word across Tuscany. To stay in spiritual prime, he whipped himself three times daily till he bled or fainted. In cold weather, he recited psalms naked by an open window.

In warm weather, he threw himself into thorn bushes. Each night he slept in a nail-studded gunnysack on a table.

The year of Steno's conversion, when he was deep in his geological studies, he spent long hours discussing religion with Lavinia Arnolfini, the ascetic wife of the ambassador from Lucca. Arnolfini, who wore a hair shirt at all times, even under party dresses at state occasions, was the person who Steno said inspired him to think seriously about converting to Catholicism.

All that remained then was his emotional leap, triggered by the voice from a window. From that point on, religion was to take up an ever-increasing share of Steno's time and energy in his life until, ultimately, there would be room for nothing else.

But on the day of his conversion there was the more immediate complication of the summons from the Danish king. Catholicism was banned in Denmark—a priest caught saying mass could even be put to death. He wrote a carefully worded reply, informing the king of his new faith, and inquiring about the possibility of religious freedom.

Steno then immediately left Florence on one of his "mussel journeys" into the hills. Religion aside, he was in the midst of making his greatest scientific discovery.

LAYER BY LAYER

Errors, like straws, upon the surface flow;
He who would search for pearls must dive below.

—JOHN DRYDEN, *ALL FOR LOVE* (1678)

"In various places," wrote Steno, "I have seen that the earth is composed of layers superimposed on each other at an angle to the horizon."

It is an amazing fact of the history of science that before Steno few European writers had thought this fundamental observation worth mentioning.

"The ancients did not know what is obvious to the eye of anyone anytime meat is brought to the table," said Steno on muscles. "What is most evident receives the least attention." No doubt, he added, "hidden among what I myself have discovered," there were things "simpler and more obvious than what I have seen." Artists, he said, often observed the body more accurately than did scientists.

But when it came to the earth, artists and scientists shared the same blindness. During the Renaissance, painters exalted the beauty and symmetry of the human body. They discovered nature, rendering plants and animals in precise, photographic detail. But the earth itself remained beneath their notice. With rare exceptions, the rocks, cliffs, and mountains depicted in paintings are no more than stylized blobs. Landscapes, even when more than just backdrops for portraits and religious scenes, make no kind of geologic sense.

An exception, of course, was Leonardo. In his notebooks he describes the sedimentary strata of Tuscany at great length, anticipating much of what Steno would later discover in the very same hills. In fact, Leonardo's earliest known drawing, made in the summer of 1473 when he was twenty-one, is a geologically realistic Tuscan landscape. The drawing's cliffs and rock strata are so accurately rendered that, according to a recent study, the strata in the foreground hills match up perfectly with those in the far distance.

Oddly, in Leonardo's later and more famous works, such as the *Mona Lisa* and *Madonna of the Rocks,* his landscapes are no more realistic than any other artist's. In *Madonna of the Rocks,* much praised for its naturalism and the accurate botanical details of the plants, the rocks are nondescript masses and mountains in the background are dramatic, but unrealistic, spires. It seems that Leonardo did not choose to share his geological insights with the viewing public.

The world before Leonardo did not completely ignore the earth's stratification, of course. Five hundred years earlier a mystical sect of Shiite Muslims in Syria who called themselves the "Brothers of Purity" knew full well what rock strata were all about.

The Brothers' motto, "Shun no science, scorn no book, nor cling fanatically to a single creed," made them unpopular with the local authorities, so they had to operate in secret. Still, sometime near the end of the tenth century they managed to publish an encyclopedia of knowledge, called *The Aim of the Sage.* In it was a brief dissertation on geologic lore, including the following on strata:

Know, O my Brother, that all rivers and streams arise from mountains and hills. In their flow they move towards the seas, ponds, and marshes . . . then [the mountains] break up, especially under the effect of storms, and become stones and rocks or pebbles and sand. . . . Then the seas, due to the force of their waves, the intensity of their agitation, and their spouting, deposit these sands, this clay, and these pebbles on their beds, layer upon layer, in the course of time and over the ages. Then [these layers] are heaped up, one on the other, and thus are formed and raised, at the bottom of the seas, mountains and hills in the same way as sand dunes are formed in plains and deserts by the effect of winds.

The Aim of the Sage was read throughout the medieval Islamic world, spreading as far west as Spain, where it became popular with Jewish scholars as well. But it remained unknown in Christian Europe.

Avicenna, the most famous of the medieval Islamic scholars, surely knew the Brothers' work and made similar observations of the mountains in his native Uzbekistan. "We see that some mountains appear to have been piled up layer by layer," he wrote in a book on minerals. "One layer was formed first, then, at a different period, a further layer was piled (upon the first, and so on)."

When Avicenna's book was translated into Latin, it was at first mistakenly believed to be a long-lost text of Aristotle's, and European scholars read it eagerly. Even after the confusion of authorship was cleared up, Avicenna was perhaps the most-cited authority on minerals and rocks.

But the Latin text of Avicenna was missing many of the original's key passages—including those on stratification. At the time, Aristotle's doctrine of the eternity of the world was under heavy attack from the church. One French monk who defended eternalism had to flee Paris, only to be tracked down and assassinated. The concept of mountains being built up layer by layer may simply have been too hot to handle.

Whether or not the omission was a deliberate act of censorship, or simply shoddy scholarship, it was not yet time for strata to enter the European consciousness.

Long before Steno it was known that certain things added layers as they grew. The ages of trees and of seashells, for example, could be figured by counting their annual growth bands. In Steno's *Chaos* journal he notes that from his reading he learned "in the Alps there is old ice in which you can distinguish days and years."

But Steno seems to have come to his realization about the significance of rock strata not from analogies with tree rings or snow packs. He seems to have reached it by a more complicated route, beginning with his meditations on the variety of things seemingly produced by the earth.

Besides marine fossils, other natural objects embedded in rock were "endowed with a certain shape." They resembled bodies, mineral as well as organic, that grew in "fresh waters, in the air, and in other fluids." As with fossil seashells, most writers attributed their presence in the rocks to the "power of the place."

Not every clam-shaped stone truly was the remains of an actual clam. Some similarities were indeed by chance, just as the shape of a cloud might resemble a horse. Crystals and minerals could be beautifully transparent and geometrically regular to an astonishing degree. These objects were in their way as mysterious as fossil clams.

"At length, I saw that the point had been reached where every solid naturally enclosed within a solid should be examined to determine if it was produced in the same place as it was found," he wrote.

"If we grant the earth the power to produce these bodies, we cannot deny it the ability of producing the rest."

Earlier attempts to explain glossopetrae and fossil seashells had been too piecemeal, he explained to Ferdinando in his introduction to *De solido*. And so, in order to "satisfy the laws of analysis,"

I wove and unraveled the web of this investigation many many times, and examined its individual parts until there seemed to me to be left no further difficulty in the reading of authors, nor in the objections of friends, nor in the inspection of sites, that I had not either solved, or about which I had at least decided, from what I had learned up to now, how far a solution was possible.

To explain tongue stones, Steno had taken on the problem of explaining fossil seashells and all other marine bodies found in rocks. To explain fossils, he had taken on the problem of explaining the formation of rocks. Now he would need to explain the production of natural solids in general. Furthermore, he said, "all discussion on the method of production is futile unless we have some certain knowledge of the nature of matter."

"From this it is clear how many problems must be solved to satisfy one line of inquiry."

And the problems were daunting indeed. The question of the ultimate nature of matter was one of the most hotly contested issues of the seventeenth century, in science *and* theology. Ironically, it had been linked to fossil seashells almost from the beginning, starting with the pre-Socratic philosophers, Thales and his students, in the fifth and sixth centuries, B.C.

Thales believed that water was the basic form of matter, and that everything was made of water. The solid earth was no more than a hardened crust formed over a primeval ocean. Earthquakes resulted from shifting of this crust on the water below. In his known writings, Thales never mentions seashells as evidence for this theory (in fact no surviving text explains any of the reasoning behind it), but as a well-traveled cosmopolitan he surely knew about them. Marine fossils abound in the geologically active eastern Mediterranean where he lived.

Thales's successors, such as the satiric poet Xenophanes of Colophon, did write explicitly about fossil seashells. Seashells in mountain rocks were evidence that water could transform into rock and back again. Matter existed in two fundamental phases: water and earth.

Later philosophers added air and fire to the mix. By the time Aristotle began his project of organizing all scientific knowledge, the four-element theory of matter—air, fire, water, earth—was fully entrenched in philosophical thinking and would remain so for almost two millennia.

But the Renaissance had seen the emergence of rival theories. A Flemish chemist named Jan Baptiste van Helmont grew a willow tree in a pot, adding nothing but water for five years. The tree's increase in mass had to come from the water. Deciding Thales had been right all along, van Helmont wrote that no solid could be generated "but by a getting of the water with childe." To alchemists, all solid matter was a combination of three "principles," namely mercury, sulfur, and salt, roughly corresponding to Aristotle's water, fire, and earth, respectively.

Most controversial, however, was the revival of the old Greek theory of atoms.

Atomism was reborn in Italy during the heyday of Renaissance humanism, when scores of ancient classical texts were rescued from oblivion by scholars, and published for the newly literate middle class.

The vehicle was a poem, *On the Nature of Things*, by the Roman Lucretius.

The atomism of Lucretius was as daring and scandalous an idea as there could be, because at bottom it denied that God or any other organized force was in control of the universe. Everything was the result of random collisions and combinations of atoms. There were no supernatural forces acting in the world, just the physical processes of nature. When the poem was written, its intent was to comfort an overwrought Roman populace at a time when superstitions and fears of demons and sorcery were rife.

That was not how it was initially received in Renaissance Europe, however. Before it was tamed and Christianized, by putting the motion of each atom under God's control and identifying atoms as the "dust" from which Adam was created, it was seen as irredeemably atheistic and dangerous to Christian morals.

Atomism presented the Catholic Church with a special problem. The Catholic rite of the Eucharist, in which bread and wine are miraculously transformed into the flesh and blood of Christ, was officially explained by the Church in Aristotelian terms: The "substance" of the bread and wine were replaced by that of Christ. The external appearances of the bread and wine, which remained the same, were "accidents." As with fossil stones, their visible form did not represent their true essence and therefore could be disregarded.

Reconciling atomism with the Church's doctrine of transubstantiation was not so easy, however. If atoms were immutable entities, how could such a transformation happen? As the Eucharist was one of the Church's primary articles of faith, one that was attacked by its Protestant enemies, the "atomistic heresy" was considered more dangerous, and more vigorously repressed, than even Copernicanism, which threatened no explicit Church tenet. According to some historians, Galileo, who was an atomist as well as a Copernican, got off easy.

At Ferdinando and Leopoldo's Cimento, the matter question was at the center of many of their experiments. Most of the scientists hewed to the "corpuscular" theory of matter, which was essentially atomism stripped of its theological and metaphysical difficulties: Matter consisted of small particles, which combined together in different ways to make material objects. They didn't care to speculate on whether the particles were divisible or indivisible, came in many varieties or few, or were separated by vacuum or a matrix of still-smaller particles. And they especially did not speculate on what happened to the corpuscles of bread and wine during mass.

Instead, they addressed their experiments to more mundane transformations of matter—the expansion of solids by heat and the freezing of liquids by cold. It seemed plausible that heat and coldness were material substances, "corpuscles of fire" that caused expansion by prying apart the corpuscles of matter they entered, and gluelike "atoms of cold" that solidified liquids by immobilizing their corpuscles. Water, which was unusual in that it expanded when it froze, was especially interesting. Leopoldo directed these experiments himself, personally sitting outside in the cold on many winter nights—occasionally inviting dinner guests at the palace to observe.

Steno, who joined in some of Leopoldo's nighttime experiments, also leaned toward the corpuscular theory of matter. In Copenhagen, he had read the corpuscular explanations of snowflakes' six-sided symmetry by Kepler and Descartes. Stacked like oranges, corpuscles naturally arrayed themselves into hexagonal patterns. It was a straightforward matter of solid geometry. Kircher, who he also read, attributed the geometry of snowflakes and other crystals to the same magnetic and plastic forces he used to explain organic fossils.

Kircher held to the traditional Aristotelian view that all matter—even rocks—was not only alive but possessed a kind of intelligence. How else would a dropped stone know which way to fall? The animal, vegetable, and mineral kingdoms all possessed life, differing only in the degree of its development. Corals growing in the sea—part vegetable and part mineral—were proof of the fuzziness of the distinction between organic and inorganic matter. *Lapides sui generis*—self-generated stones—not only grew, but gave birth, had digestive organs, and came in male and female genders. The cumulative growth of stones could build mountains.

This seemed to fit with everyday experience. Farmers found the stones in their fields to be endlessly abundant. Every year the frost heaved up a new batch; by all appearances they grew in the soil like weeds. Miners had long believed that minerals grew like plants in the interstices of the earth. In mines, it was said, "the Iron breedes continually as fast they can worke it."

But in his visits to mines, Steno had seen no such thing. As an anatomist he knew how elaborate were the body's organs that maintained its life. There was a fundamental difference between organic and inorganic matter. Minerals and rocks did not grow like plants.

To many people, including Steno, the animistic view of nature was unacceptable on religious grounds. Redi began his tract on spontaneous generation by saying that after the first week God stopped creating, therefore every living thing must be a direct descendant from that first Creation.

Attributing too much creative power to nature led to pantheism and a rejection of God the Creator. Nature viewed as self-existent and self-maintaining was just a stone's throw away from a nature that was self-creating—that is, one in which there was no need for God.

Steno and the other scientists who embraced corpuscular physics and, along with it, the mechanical philosophy, were trying make sure that God wasn't squeezed out of natural philosophy by nature. Matter, said the mechanical philosophy, was dead, inert. It could do nothing on its own. It required an outside power, and that power was God. The mechanical philosophy made the division between the material and the spiritual world absolute. There was God, the angels, and the human soul on the spiritual side. Everything else was just dumb matter.

The question facing Steno was how dumb matter could arrange itself into all the intricate natural objects found inside and outside of rocks.

According to corpuscular theory, the difference between solids and liquids was that in solids the particles are closely bound together, whereas in fluids (liquids and gases) they could freely move about.

Solids grow from fluids—this was a common denominator of both crystals and living things. The only way a solid grows is by the addition of new particles, and the new particles have to be brought to it via a fluid. Because the solid particles can't penetrate one another, this means that the particles are added to the surface. They therefore attach to the surface a layer at a time. Steno decided that to understand how different kinds of solids grew, he would need to understand how the layers were added in each case.

Now at Florence, in the laboratories of the Cimento, Steno was studying mineral crystals and mollusk shells. One day the grand duke gave him a large pearl to crack open and study. At its core was a tiny black grain of sand. Gleaming white layers had grown around the grain, one atop the other, to produce the pearl. Also in the grand duke's collection were laminated stones and agates that showed a similar concentric pattern, but lacked a nucleus. They had formed in small voids inside the earth. A layer of mineral had first formed on the inside surface of the void, later layers had accreted, building inward until they filled the void.

For the pearl and the laminated stone, the pattern was similar, concentric layers, but the history of growth was different. For the pearl, growing outward, the inner layer was the oldest. For the laminated stones, growing inward, the outer layer was the oldest.

It was not enough for Steno, or anyone else, to notice the layering of the earth's bedrock. He had to grasp that the layering represented a sequence, and he had to understand the direction in which the sequence ran. The sequences of layers comprised a chronology. Neither Leonardo nor the Brothers of Purity saw this as clearly as did Steno.

The idea is simple, but it was not obvious.

Before Steno, René Descartes had made his own attempt to outline the history of the earth. Characteristically, he had based his theory on deduction from first principles rather than from the study of the earth's crust itself. He, too, imagined a layered earth, but the layers were remnants of the planet's first congealing out of primordial matter. Descartes reasoned that different kinds of matter (consisting of differently shaped "corpuscles") would separate from one another, in a fashion similar to the separation of oil and vinegar in a bottle of salad dressing.

Other theorists, who like Descartes hypothesized that the earth had initially been hot and molten, thought that as the earth cooled its layers solidified from the outside in, making deeper layers "younger" than those closer to the surface.

An English theorist named John Strachey, writing some fifty years *after* Steno, noticed that the strata in his part of England were mostly inclined, not horizontal. He denied that they were sediments and that they had formed in any particular sequence. He imagined that the earth's layers all spiraled into the center, like a pinwheel. An originally molten earth, spinning as it solidified from the outside in, would naturally form this structure, said Strachey.

But Steno knew the world was more like a pearl than a pinwheel.

He does not record when or where he made the crucial connection, or what it was that led him to his insight, but by the spring of 1668, everything seemed to fit together. After excursions to the hilltop city of Volterra and the mines on the island of Elba, he wrote enthusiastically to Magalotti, "Everything I saw confirmed my opinion, or rather the opinion of the Ancients which I defended in my last treatise" regarding "the origin of mussels, shellfish, and glossopetrae found in mountains."

In June, he broke off his research and begin to write. Perhaps he received some communication from Denmark that a new royal summons (with the necessary religious immunity) was imminent. Although Steno's research was not yet complete, he had enough confidence in his ideas to write a short abstract, a prodromus of the full dissertation he had promised Grand Duke Ferdinando.

He worked quickly. In a few weeks, *De solido intra solidum naturaliter contento dissertationis prodromus,* "Prodromus to a Dissertation on Solids Naturally Enclosed in Solids," was finished and ready for the censors.

10

DE SOLIDO

See, I cast the die, and I write the book. Whether it is to be read by the people of the present or of the future makes no difference: let it await its reader for a hundred years, if God himself has stood ready for six thousand years for one to study him.

—JOHANNES KEPLER, *HARMONICES MUNDI* (1619)

The ancient puzzle of fossil seashells was solved at last, Steno announced in the opening pages of *De solido*. Or to be precise, as he hastened to point out, it had been solved *again*.

The question was "old, pleasant, and useful." The ancients had never hesitated to identify the fossil seashells as true seashells. Later investigators departed from this view, Steno wrote, and while "many excellent books by a large number of writers" had been published, the controversy had never been settled. In fact, as with other scientific issues, "the doubts multiply with the number of writers."

For all his deference to the ancients, Steno had taken an important step beyond what they, or any writer since them had taken. The old writers who recognized the identity of fossil shells, and had accepted Aristotle's theory that the face of the earth had a past marked by dramatic changes, also had accepted his assumption that the earth's past was chaotic and unintelligible. They had been satisfied to simply note the shells in the mountains and hold forth on the world's mutability.

But Steno was the first to assert that the world's history might be recoverable from the rocks and to take it upon himself to unravel that history. And he did so in a way remarkable in its simplicity, based on elementary geometry. He distilled his principles from the most easily understood geometrical relationships: above versus below, continuous versus discontinuous, tilted versus horizontal, enclosed versus enclosing.

Working under the pressure of time, what Steno wrote was intended to be just a boiled-down summary of a longer and more complete work that would follow, "therefore sometimes I refer to observations, sometimes to conclusions, whatever may seem best to indicate the chief matters as briefly and as lucidly as possible." All details of fact and logic would follow in the full dissertation.

The idea behind his approach, he said, could be summarized very briefly:

> Given a substance endowed with a certain shape, and produced according to the laws of nature, to find in the substance itself clues disclosing the place and manner of its production.

This is the sine qua non of historical science. Unless it is possible to learn about past events by studying present objects, all hopes of scientifically studying the past are futile.

Simple as it was, this statement constituted a break with previous thinking about history, which relied on written documents or human artifacts, and with emerging thinking about science, which was becoming increasingly focused on timeless laws and "real time" experimentation. Natural historians had traditionally studied objects in order to classify them. Mechanical philosophers studied them to learn how they functioned. Studying them to ascertain their history was a new concept.

The objects he was interested in, besides being contained in other solids, were those that grew from fluids. For organic beings, such as animals and plants, the fluids are the internal fluids—blood in animals and sap in plants that transport materials to the growing site. The bodies of plants and animals are full of plumbing to shunt these fluids around so that the right matter can be delivered to the right place and added to the right internal surface so that it grows properly. One result of this is that organic materials—even stony materials such as bones and mollusk shells—have a complicated and distinctive internal structure to them. "Even if marine testacea had never been observed," he concluded, "examination of the shell itself proves that these shells were once parts of animals."

For inorganic materials, the fluids are external. Crystals grow as particles out of solution and adhere to their surfaces, building up layer by layer. They do *not* grow, as many of Steno's contemporaries believed, like plants, drawing mineral "nutrients" from their roots. Nor were they produced by God at Creation, as contended by others who could not accept that the filthy earth could produce such "pellucid bodies."

Steno's fascination with crystals and their symmetry, which ten years before had sent him out into the Copenhagen streets to sketch snowflakes, also led him to a discovery that later formed the basis of crystallography, the science of crystals.

Like snowflakes, perfect crystals of quartz have a six-fold symmetry. Steno noticed, however, that even in crystals that had grown unevenly so that the symmetry was distorted, the angles between the crystal faces stayed constant. Steno was unable to explain this geometric "law" of crystal growth, and could not have guessed its significance. Only a century later did scientists begin to realize that the characteristic "inter-facial" angles of mineral crystals were key evidence of their internal atomic structures.

Steno's seminal insights on minerals and their growth were important steps in creating the nascent science of geology—minerals are, after all, the building blocks of rocks and therefore the earth's crust. But when Steno set out to understand the "place and manner" of a natural solid's production, he had more in mind than just the individual object's growth from a fluid. For a solid inside a solid, the full story involved the enclosing solid, too. Which came first? Steno found a simple criterion:

> *If a solid body is enclosed on all sides by another solid body, the first*
> *of the two to harden was that one which, when both touch, transferred*
> *its own surface characteristics to the surface of the other.*

This principle established the relative sequence of hardening between any body and its surrounding context, and between a body's various parts.

By this reasoning, a cockleshell found in sandstone, impressing the texture of its ribs on the surrounding stone, grew before the sandstone solidified. In contrast, an oyster nestled in a nook of a sandstone boulder, its shell conforming to the shape of the nook, grew after the sandstone was already solid. By the same logic, a quartz pebble hardened before the sedimentary deposit that contained it, but a quartz vein following a fracture through the same deposit solidified afterward.

Much of *De solido* surveys the various natural objects found in the earth and according to these principles, sorts them into the categories of whether they grow within the rocks or outside, whether they existed from the "beginning of things" or not.

Some objects revealed more complicated histories. One shell he describes had been petrified, its interior filled with a "marble incrustation." It had later been partially destroyed, and the broken surface was covered by barnacles, which were also fossilized.

To Steno, this implied the following story: The animal that had made the shell died and was buried in sediments, where the shell became fossilized. The sea left and the sediments, now dry land, were eroded and the fossil shell was exhumed and carried down to the sea. Along the way it had become partially destroyed. Once back in the sea, however, it had lain on the sea bottom for a while and become encrusted by barnacles. The shell was then buried in a new deposit, which, after another retreat of the sea, yielded the fossil.

This single fossil, easily hefted in the hand, spoke of not just one inundation in an ancient sea, but *two*. It also implied not only the advance and retreat of ancient seas, but the erosion of a vanished landscape, the one the shell had tumbled over on its way back to the sea.

Steno did not limit himself to telling isolated stories about individual seashells, however. To fit fossils, rocks, and strata into a single narrative, he proposed a set of principles, three of which have been immortalized as "Steno's Principles." These are taught to every student of geology, and form the basis of all investigations into the geologic past.

The most famous is his "principle of superposition." Given sedimentary layers arranged one on top of another, the layer on the bottom was deposited first and the one on top was deposited last.

"When the lowest stratum was being formed, none of the upper strata existed."

So sedimentary strata are formed *in sequence*. They record a series of events, they reveal the steps by which they were formed. The layer at the bottom is the oldest, the layer at the top is the youngest.

The principle, as an abstract geometric principle, is common sense. But the key is realizing that this particular principle of common sense applies to the rocks of a mountainside. To understand that a stack of objects gets piled up from bottom to top required no great intellectual leap. But understanding that layers of bedrock contained a *narrative*, that it made sense to speak of one rock being "older" than another, was Steno's critical breakthrough.

Steno's principle of superposition marks the beginning of stratigraphy, the science of geologic strata. The first task of the stratigrapher is to figure out the arrangement of geologic strata within the earth. From there, he or she reconstructs geological history.

When thinking about geologic strata, the image that comes most readily to mind for most people is the spectacular stratigraphy of the Grand Canyon—a huge layer-cake of rock formations laid out in a single, breathtaking vista. The canyon is indeed a wonder of geology, but that is the point: Only rarely is the underlying geology of a landscape so plainly and unambiguously visible.

Most landscapes—at least most where people are apt to live—are covered by a blanket of vegetation, with perhaps a few rocky crags poking out here and there. And most of those rugged landscapes where rocks are exposed in abundance—seashell-bearing mountain ranges for instance—got that way because their bedrock was at some time crumpled and pushed upward by geologic forces. The layer cake has been mashed and tilted and otherwise distorted, and erosion has taken large bites out of it, too.

So to help sort through this mess, *De solido* serves up two more principles no more complicated than superposition.

Water is the source of sediments, said Steno. One sure thing about water is that, no matter what the shape of its container or how it is tilted, its upper surface always lies parallel to the horizon. Obeying gravity, the waters of every sea or lake will inevitably spread out evenly to fill their basin.

So, too, when the water's load of sediment sprinkles down to make a layer, it will blanket irregularities on the bottom, and its upper surface will be smooth and roughly horizontal. As the sediment accumulates, these layers will also be horizontal.

This means that no matter what a stratum's orientation is now, if it was born as a deposit from water, it was originally horizontal. Any tilting or folding is a consequence of later events. Geologists call this Steno's principle of "original horizontality."

The earth's strata are often clearly *not* horizontal. In mountain ranges where folding has been intense, it's not rare to find sedimentary beds that have been tilted to the vertical—and beyond. In these places, the strata may not resemble water-laid deposits much at all. It takes some imagination to mentally straighten out the folds and see the strata as originally horizontal layers.

Of course, to do this, an investigator needs to know which way was originally "up." If the beds are vertical, how do you "read" the sequence of superposition? Left to right or right to left? If they have been overturned so that the sequence is upside down, how could you tell?

Steno's observations on how water lays down sediments help here, too. When the sediment carried by water begins to settle out— as, say, the river current slackens or the storm abates—it's always the largest and heaviest grains that settle first. They need the strongest turbulence to stay in the mix. Progressively smaller grains drop out in sequence. Fine clay, which barely needs any agitation to stay suspended and takes a long time to settle, will be the very last.

As a result, the particles within a sedimentary layer are commonly sorted by size, the coarsest on the bottom and the finest at the top. Thanks to gravity, these beds are marked "this side up."

After superposition and original horizontality, the third and final principle gleaned from *De solido* is that water lays down sediments as laterally continuous sheets, ending only at the edge of their basin. "Wherever bared edges of strata are seen," Steno wrote, "either a continuation of that same strata must be looked for or another solid substance must be found that kept the material of the strata from being dispersed." Corresponding rock layers on opposite sides of a valley were formerly connected as single, continuous layers. This is Steno's principle of "lateral continuity."

Just as tilted strata imply past movements of the earth's crust, "bared edges" of strata also imply a history. Some previous landscape was carved into the present one; the valley separating the strata had not existed. What was now a hill had formerly been a basin.

To study a solid such as a fossil, crystal, or rock stratum for "clues disclosing the place and manner of its production," was a kind of science very different from the controlled experimentation of Redi or the abstract deduction of Descartes. The thinking was more like that of a detective, who might use the same kind of reasoning at a crime scene, looking for clues to build a case about what happened and when.

Ancient seas left their footprints in strata in the form of tongue stones and seashells. Ancient river floods left their fingerprints as fossilized leaves and petrified wood.

Fixing a time for these events was harder. Large fossilized bones found near Arezzo, upriver from Florence, might be the remains of the elephants brought across the mountains by Hannibal in 218 B.C., Steno thought. His biological identification was close enough, but his dates were way off. The deposits date from the Pleistocene, popularly known as the Ice Age, and are hundreds of thousands of years old.

But for now, such details weren't important. Steno's chronology was a matter of sequence, not dates or numbers of years. There was no telling how long ago a given layer of rock was laid down, or how long it took to be laid down. But Steno's principle of superposition made it beyond dispute that it was laid down after the layer below it and before the layer above it. Scientific geology began with the "which came first" question, rather than the "how long ago." Questions of relative time were answerable with the information at hand, those of absolute time were not.

To Steno, separating answerable from unanswerable questions was the key to scientific investigation. Questions about the past had always been assumed to be unanswerable. To show that answers could be found in "natural solids" of all kinds, he labored to come up with principles so simple and clear that "no school of philosophers may be left in doubt and dispute about them."

He would be disappointed. Speaking more prophetically than he could have known, Steno said that consensus would be impossible so long as basic questions of evidence remained in dispute. Some spurned experiments, others spurned theory. Others speculated with abandon, "they being of the opinion that all those things are true that seem to them admirable and ingenious."

Ultimately, Steno's achievement in *De solido* was not just that he proposed a new, and correct, theory of fossils. As he himself pointed out, writers more than a thousand years earlier had said essentially the same thing. Nor was it simply that he presented a new and correct interpretation of rock strata. It was that he drew up a blueprint for an entirely new scientific approach to nature, one that opened up the dimension of time. As Steno wrote, "from that which is perceived a definite conclusion may be drawn about what is imperceptible." From the present world one can deduce vanished worlds.

Armed with his new science, Steno was emboldened to delve deep into the past, to explore a new history of the world "not dealt with by historians and writers on things of nature."

11

THE SIX FACES OF TUSCANY

Our age cannot look back to earlier things
Except where reasoning reveals their traces.
—LUCRETIUS, *ON THE NATURE OF THINGS*

"**H**ow the present state of anything discloses the past state of the same thing," wrote Steno in the concluding section of *De solido*, "is made abundantly clear by the example of Tuscany, above all others."

By holding up Tuscany as an epitome of the earth's geologic evolution, Steno no doubt knew he was appealing to his patron's pride in his home country. Galileo had used a similar rhetorical ploy in naming the moons of Jupiter the "Medician" stars. The first new worlds to be discovered by science were given the name of Tuscany's most famous family. Now Steno claimed that the same family's home turf was a literal paragon for the evolution of the present world.

Such geologic chauvinism was not limited to the quaint old days of the seventeenth century. British geologists in the early nineteenth century who delineated the geologic time periods that today make up the standard geologic timescale took immense pride that their island served as an exemplar for our planet's past. Dinosaurs from the American west have sometimes also been pressed into patriotic duty by those who see these powerful (but extinct) creatures as symbols of American might.

But if Steno was laying it on a little thick to appease the grand duke, he was right that Tuscany offered ample opportunities for geologic investigation. He was lucky, in a way. The landscape over which he traveled revealed something of the geological past, but not too much. Had Steno known the true scale of his undertaking, he might have hesitated to try to unravel the geological story of the Tuscan rocks.

The Italian peninsula lies at the juncture of several tectonic plates. Multiple waves of mountain building have deformed and redeformed its bedrock over hundreds of millions of years. During these convulsions, the strata making up the Appenine mountain chain were squashed, folded, and in some places even flopped upside down. Later geologists have often despaired at deciphering the complexities of Tuscan geology.

Fortunately, Steno did not attempt to describe it in any detail. So far as is known, he did not do what generally constitutes the first phase of the geologic exploration of a region: He did not make a map. He did not sketch out the areas in Tuscany where different sorts of bedrock can be found, or, if he did, the map did not survive among his papers.

What has survived is a highly schematic diagram of the geologic structure of Tuscany, viewed in cross section. Self-consciously geometric like his muscle diagrams, it is the first geologic cross section ever made.

It is actually a sequence of six diagrams, each representing a different stage, "*facies*" of Tuscany's geologic evolution. They fall into

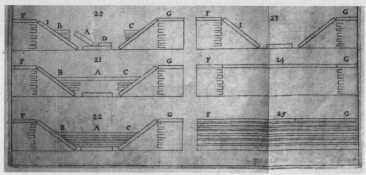

Steno's geologic cross-section of Tuscany, showing the six different stages of its development. The sequence begins with diagram 25 (lower right) and ends with diagram 20 (upper left). Steno envisioned two cycles; in each, sedimentary strata are laid down, then undermined by the growth of caverns, and finally collapse to create mountains and valleys. Solid lines represent rocky strata, and dotted lines represent sandy strata, "although various strata of stones and clay are mixed with them." Nicolaus Steno, *De solido* (Florence, 1669). Courtesy of History of Science Collections, University of Oklahoma Libraries.

two cycles of three stages: the land is flooded, drained, and then buckled into mountains.

Where today "rivers, swamps, sunken plains, precipices, and inclined planes" are found, Steno begins, "everything was once level."

This primordial surface was covered by water, laying down the first series of horizontal sedimentary layers. Next the water withdraws, exposing a flat, dry plain. By now the strata have hardened to rock, but they are undermined "either by the force of fires or waters," which progressively eat away at them from below. When the underground cavities grow so huge that the earth collapses, the formerly smooth landscape becomes transformed into a rugged terrain of mountains and valleys. Steno's diagram shows the uppermost strata steeply tilted where they have flopped down like trapdoors into the caverns.

The next cycle repeats the events of the first, although this time when the land is flooded only the valleys receive sediments, which are mostly soft and sandy, not rocky as before. After the sea withdraws, these strata also develop caverns, which collapse, creating another generation of hills and valleys.

Steno's explanation for the irregularities of the landscape— stupendous cave-ins—seems odd today, but would have been conventional at the time. The idea of a cavernous earth had been around since Plato, and Descartes had also hypothesized that mountains were created by a collapsing crust.* No doubt Steno was impressed by the caverns he visited in Tuscany, and he probably assumed they were deeper and more extensive than they really are.

Steno's geology may actually have been better than scientists have previously believed. His hypothesized caverns never existed, but Italian geologists have recently pointed out that the bedrock underlying the Tuscan landscape is fractured by steeply inclined faults. Many of Tuscany's broad, deep valleys were created when, beginning about six million years ago, the earth's crust in that part of Italy was stretched and pulled apart by geologic forces, causing blocks to collapse along these faults, just as a row of books would topple if a bookend was pulled away. Afterward, these tectonic valleys were flooded by a rise in sea level, which deposited the shelly, sandy, and clay-rich sediments Steno observed. According to the Italian geologists, the structure of several valleys in Tuscany matches Steno's description surprisingly well.

Steno was certainly confident enough about his conclusions to take the next step: "What I demonstrate about Tuscany by induction from many places examined by me," he said, "so I confirm for the whole earth."

---

*The idea of a hollow earth survived even into the nineteenth century, and not only in fictional tales such as Jules Verne's *Journey to the Center of the Earth*. When in 1838 the United States government sent its first scientific expedition abroad to explore unknown polar regions, one of its stated aims was to search for passageways into the inner earth.

In saying this, Steno was sticking his neck out. Years before, when René Descartes was putting the final polish on his theory of the world's origin before he handed it to the printer, he caught word of Galileo's trial. That was enough to change his mind about publishing. When the theory did come out after his death, it was peppered with disclaimers. The theory was only an exercise to show how the earth and stars *could have arisen* even though, he said, "we know perfectly well they did not originate in this way."

Steno made no such excuses, but he was clearly worried some people might view his ideas on the geologic changes in the earth as being impious. "Lest anyone be afraid of the danger of novelty" (that is, a new and unapproved interpretation of Scripture), he wrote, "I set down briefly the agreement between Nature and Scripture."

It is not clear who he saw as his potential critics or how concerned he was personally with adhering closely to Scripture. Was he concerned about the local Catholic authorities, who presumably would not hesitate to squelch any deviation from the received interpretation of Genesis? Or was he concerned about his bid for a job and religious freedom in his Protestant homeland, where biblical literalism was more strictly observed than in Catholic Italy?

Regardless of the religious situation, the Bible was universally considered in the West as the most reliable of all historical documents. How could a fastidious researcher like Steno neglect to corroborate his results against it?

And when he did, Steno was pleased to find everything fit together splendidly.

The key events in his geologic history were the two inundations—one to lay down the hard, rocky strata of the high mountains, and the other to lay down the sandy strata of the foothills. Happily, Scripture recorded two watery phases as well.

The first came at the very beginning of the Creation week, starting on the first day when "the spirit of God moved on the face of the waters," and ending on the third when God gathered the waters to let dry land appear.

This fit perfectly with Steno's geology. "That the fluid covered everything, is proved conclusively by the strata of the higher mountains," he wrote. In these rocks he had found no fossils, so they must have been deposited "at a time when animals and plants had not yet appeared." This also fit with Scripture: Animals and plants weren't created until after dry land had emerged. Marine life wasn't created until day five. And, finally: "the similarity in materials and outlines of strata from different mountains that are widely separated proves indeed that the fluid was universal."

The second inundation might seem more problematic, he said, "though in truth it is not difficult." "Nature does not contradict what Scripture determines about how high the sea was." Oddly enough, he does not explicitly mention Noah here, but instead argues that because waters covered the earth at "at the beginning of things," they could conceivably do it again "since change is indeed continual in the things of nature, but nothing in nature is totally destroyed." Elsewhere in *De solido* he implies that at least some areas have been flooded multiple times, not just the single time of Noah. Perhaps this is part of the continuity he mentions.

As to how the waters rose, he offers two possibilities. One is an old theory first proposed by the fourteenth-century Scholastic philosopher Jean Buridan.

Buridan believed that because parts of the earth's interior were cavernous, its center of gravity did not always line up with its center of figure. That is, it was basically lopsided. The seas mainly occupied the low part of the earth, but because the cavities deep in the earth occasionally collapsed, and because debris from the high land was continually being washed into the sea, over time, the earth's center of mass would shift. As a result, different areas of the earth would be covered by the sea at different times.

Alternatively, Steno thought that maybe there were reservoirs of water deep in the earth that occasionally boiled over from the earth's internal heat. Shooting out from the earth, this water would cover the land for a time and afterward drain back into the depths.*

Steno was less interested in discovering how it happened than in demonstrating that it did happen. And his comparing geologic strata to the Bible was not simply a pious gesture. Once he realized that strata were a record of time, he tried to fit them into a bigger picture, a narrative. Others who had believed in the organic explanation for fossils had invoked change, but had not put these changes into any kind of framework. As often as not, they had assumed that the inundations that left the fossils behind had been just local inundations.

Steno naively assumed that the strata were evidence of global events with uniform effects, deposited during periods when all land was totally submerged. It was a mistaken, but fruitful assumption. It implied the possibility of finding global, as opposed to just local, history in the rocks. It was the first step in creating a global time scale.

The basic idea of the geologic timescale is that layers of rock correspond with global time episodes—even if local events are different. For any layer in one place there are likely to be others elsewhere in the world that overlap in time with it.

Today we give these episodes names such as Devonian, Jurassic, and Cretaceous based on the bedrock of specific places—Devon in England, the Jura mountains in France and Switzerland, and the Chalk lands of Western Europe. They are local names, but they refer to global tracts of time.

We sort out the rock layers in different places by the fossils they contain. The evolutionary cast of characters has continually changed through geologic time, and rocks of similar age have similar collections of fossils. But in Steno's time fossil species were not well-enough known to do this. Scientists had barely begun to classify species of *living* marine creatures and map their geographic distributions. The state of knowledge for fossils was even worse.

In *De solido,* Steno did not give any information on what fossils and minerals he found where or the arrangement of strata at specific localities. That was in the dissertation itself, to which he would soon add "the finishing touch" after his upcoming journey. Then he would give "clear proof of the many changes that have occurred in Tuscany."

It was not possible to fix the time of these changes, he said, but that they happened was certain. In his dissertation he would prove his new history of the world, "so that no one will be left in any doubt."

---

*An explanation that Steno seems to have rejected out of hand was the strange but enduring theory that the level of the ocean was actually higher than that of the continents, poised to flood the land but miraculously held back by God's will. This seemed to be implied by certain passages in the Bible (as in Psalm 33: "He gathereth the waters of the sea together as an heap") and by the ability of many people to convince themselves while staring at the ocean that the water bulges upward toward the horizon. One of these people was Christopher Columbus. When he set sail into the Atlantic he thought he was traveling *uphill.*

12

SHELL GAME

How trivial a thing a rotten shell may appear to some, yet these monuments
of nature are more certain tokens of antiquity than coins or medals . . .

—ROBERT HOOKE,
LECTURES ON EARTHQUAKES (1705)

With *De solido*, Steno opened up unexplored scientific territory. Despite his previous doubts, Steno was now certain he had solved at last the fossil problem. He had also set down new scientific principles by which the ancient events of the world's past, unchronicled by human authors, could be discovered. The full dissertation, once it was completed, would substantiate all the points he had so far only outlined.

But Steno's own immediate future was far less certain. The situation in Denmark was still unresolved. As for *De solido*, there was no predicting how it would be received by the scientific community. "Travelers into unknown realms frequently find, as they hasten on over rough mountain paths toward a summit city," he wrote in its preface, "that it seems very near to them when they first descry it, whereas manifold turnings may wear even their hope to weariness."

It was his metaphor for the halting progress of experimental science.

It was also an almost literal description of the next few years of his scientific career.

Steno finished the *De solido* manuscript midsummer 1668 and delivered it to the church's vicar general in Florence for permission to have it published. With its explicit references to Scripture, Steno's geological manuscript would seem to have invited especially close scrutiny by the church. In fact, it was sent to the two most sympathetic censors imaginable—Steno's friends Viviani and Redi. Not surprisingly, they gave both author and manuscript high marks for honesty, piety, and scientific merit. Viviani quickly returned the manuscript. For some unknown reason, Redi procrastinated, not returning it to the Holy Office in Florence until December, when it was promptly given an imprimatur.

Steno was under pressure to leave Florence; he had already quit his rooms at the Palazzo Vecchio back in July. His travel plans on hold, he lingered through the fall, waiting for Redi. Finally in November he could wait no longer and left. Viviani would see to the final details of getting *De solido* ready for publication.

Here, Steno's movements become difficult to follow. Ostensibly, he was on his way to Denmark to report for duty at the Danish king's court, but he clearly had other things on his agenda as well. His route was anything but direct, first heading south to Rome and Naples, touring southern Italy, then spending a month with Marcello Malpighi before heading north into the Alps. As much of a rush as he had been in to depart for Copenhagen, he was evidently in no rush to actually arrive there.

Steno was likely biding his time, waiting for further word from Copenhagen as friends negotiated the terms of his appointment at the Danish court. As a Catholic, Steno would have been barred from any official position at the university, and the post of "Royal Physician" would have been similarly problematic, though anything less would have been beneath his status as an internationally acclaimed scientist. Steno was also plainly worried that he would be denied the right to practice his newly confessed faith in that Lutheran stronghold.

The question turned out to be moot, at least for the time being, because Steno never reached Copenhagen. He traveled for twenty months, covering nearly four thousand miles, looping through not only Italy, but the Swiss and Austrian Alps, passing through Vienna, and meandering as far afield as northern Hungary.

But the closest he got to home was Amsterdam, where, in February 1670, he got word that King Frederick was dead. The deal, whatever had been or was being negotiated, was now off.

Steno may have grieved over the loss of his king, but he was more than likely relieved to be off the hook regarding the anatomy job in Denmark.

He had by no means given up on anatomy. During his travels he had done a number of dissections for the entertainment and edification of his hosts along the way, mostly friends or relations of the Medici brothers. But now he was "fully devoted" to his geological studies, collecting minerals and fossils at every opportunity, sending samples back to Florence to add to the Medici collections. If his long journey had begun as a trek toward a job in Denmark, it became over the months a barnstorming tour of the geological highlights of Europe

He had seen the famous Mount Vesuvius in the south of Italy; in the Alps he had seen high peaks and fantastically contorted strata, in the famous mines of Germany and Hungary he had seen rich deposits of minerals. He spent almost six months in Germany, an interval of time about which almost nothing is known. The only existing records of this leg of his journey are the fossil and mineral specimens he sent back to Florence to add to the grand duke's collection.

After learning of the death of the Danish king, Steno found himself at loose ends in Amsterdam. He had been in Holland almost six months, staying at the home of an agent of the Medici family. What he planned to do next, or even if he had any plans, historians have no clue.

If he did have plans, however, they were abruptly discarded when word came from Florence that Ferdinando had fallen deathly ill. With funds sent by Leopoldo, Steno made hasty preparations to return to Tuscany.

In the meantime, *De solido* had finally been published in the spring of 1669, while Steno was exploring the Austrian Alps.

It immediately went on sale at bookstores in Italy. Steno was sent copies from Florence, which he distributed to his hosts and friends as he traveled from place to place. But in general, the slim, enigmatically titled volume was slow to find its way into the hands of readers, and slow to generate much of a reaction.

It was not until 1671, two years after its publication, that the first copies of *De solido* arrived in England, where it came to the attention of the Royal Society of London for the Improving of Natural Knowledge. That was when things started to happen.

The Royal Society of London, as it is known today, is the most venerable and surely the most prestigious scientific society in the world. If all the world's books save the Society's journal, *Philosophical Transactions*, were destroyed, said member Thomas Henry Huxley in 1866, "it is safe to say that the foundations of physical science would remain unshaken, and that the vast intellectual progress of the last two centuries would be largely, though incompletely, recorded."

But when Steno's *De solido* was published, the Royal Society was a fledgling band of gentleman virtuosos, scholarly amateurs who had dedicated themselves to experimental science, yet were still feeling their way toward what that really meant.

Among its founding members were the chemist Robert Boyle, the architect Christopher Wren, and John Wallis, England's leading mathematician until he was overshadowed by his student, Isaac Newton. Taking as their motto the Latin phrase *Nullis in Verba,* "Not in Words," meaning "don't take anyone's word, see for yourself," they modeled their science after the experiments at Ferdinando and Leopoldo's Cimento. In his youth, Boyle had visited the Medici court on a tour of Italy, and though the dirt, brothels, and rampant papistry offended his Puritan sensibilities, he was galvanized by the science flourishing under Ferdinando.

Persuading the new King Charles II to grant them a charter in 1662, the virtuosos may have hoped for the same kind of support Ferdinando gave his scientists. But Charles had other fish to fry, and no money to throw away on scientific trifles. Even after they had elected him an honorary fellow, funds had to come from their own pockets. Experimental science was a rich man's pastime, though the scientists dedicated their efforts not to financial gain, but "All for the Glory of God, the Honor and Advantage of these Kingdoms, and the Universal Good of Mankind."

Despite these lofty ideals, the whole enterprise of science did not get much respect in England. In Ferdinando's Tuscany, every prince and nobleman at least feigned an interest in the latest scientific trends. Not so in England. The gentleman savants were more apt to be mocked than emulated by London's smart set, who saw them as self-important dilettantes obsessed with trivia. In his *History of the Royal Society,* member Thomas Sprat vigorously denied the charge, although the compendious 438-page tome, published in 1667 when the Society had been in existence barely five years, probably did more to confirm than refute the critics.

Visiting the Royal Society in London, Lorenzo Magalotti wrote home to now-Cardinal Leopoldo. "I could never tell Your Highness how prejudicial it is for a man of fashion from that side of the mountains to pass for a philosopher and mathematician," he said. "The ladies at once believe that he must be enamored of the moon, or Venus, or some silly thing like that." Nor were the scientists any better regarded by their royal patron. It was Charles's habit, reported Magalotti, to refer to them as "my ferrets."

Sprat was probably correct when he wrote in his *History:*

I confess I believe that New Philosophy *need not (as Caesar) fear the pale, or the melancholy, as much as the humorous, and merry: For they perhaps by making it ridiculous, because it is* new, *and because they themselves are unwilling to take pains about it, may do it more injury than all the Arguments of our severe and frowning and dogmatical* Adversaries.

Still, with the Cimento now defunct, the Royal Society was now the world's premiere scientific forum. As a way of raising money, Henry Oldenburg, the Society's secretary, had begun publishing letters sent to him by his wide network of scientific correspondents as *The Philosophical Transactions of the Royal Society.* Reviews of new books and announcements of new discoveries could be quickly disseminated among interested parties.

Ever since his first publications on glands, Steno's meteoric career had been closely followed in the pages of the *Philosophical Transactions.* Several Society members knew Steno personally from encounters in

Holland and France. Steno's letters to one of them, William Croone, were read aloud at the Society's weekly meetings. With Steno's growing fame, each new publication of his was an eagerly anticipated event.

When the first copies of *De solido* arrived in London, Oldenburg immediately published a review, and took the further step of translating the entire book into English. So there would be no confusion about the book's subject matter, he added his own explanatory subtitle:

> *Laying a Foundation for the Rendering a Rational Accompt both of the Frame and the Several Changes of the Masse of the EARTH, as also of the various Productions in the same*

A dissertation on the productions of the earth might have seemed to be a departure for the great anatomist Steno, but Oldenburg knew it would catch the attention of the Society membership. Stones and petrifaction had been topics of interest, almost an obsession, from the very first meetings.

"There happened an excellent good discourse about petrefaction," recounted member Robert Hooke to Boyle in 1663,

> . . . *several instances were given about the growing of stones: some, that were included in glass viols; others, that lay upon the pasture ground; others, that lay in gravel walks; which was known by putting a stone in at the mouth of a glass viol, through which, after a little time, it would by no means pass. Next, the story of a field's being filled with stones every third year, was confirmed by some instances. And that the stones in gravel walks grows greater, had been proved by sifting those walks over again, which had formerly passed all through the sieve, and finding abundance of stones too big to pass through the*

second time. Upon this, mention was made of the production of stones or lapidious concretions in the bodies of animals, and abundance of very strange instances were alledged of the finding of stones in several parts of a man's body. Mr. PELL and some others mentioned to have read somewhere an observation, there were more such concretions taken from one man, than the weight of his whole body amounted to.

Petrified shells were regularly brought in to be examined at meetings, members voicing their opinions, theories, and various shots in the dark. During the same lively meeting,

Mr. PALMER related a story of a French physician (whose name I have forgot) who landing sick at Dover, and taking a glister, voided an incredible number of small and great cockle-shells. The matter of fact was confirmed by very many of the Society, who had either had very good relation of it, or seen some of the shells. Dr. CHARLTON added, that they had lain a good while upon sea, and fed upon nothing but cheese (made of the milk of goats, which fed upon the mountains of Bononia which are very full of such shells) and brandy.

Their fascination with petrified objects was not idle. In fact, some members believed that it would be by solving the riddle of petrifaction that the Royal Society would answer its critics and lay to rest the charge that their experiments were esoteric dabbles. "If it lay in the power of humane skill (by the knowledge of Nature's works) to raise Petrification, or allay, or prevent it, or to order and direct it," wrote one earnest virtuoso "much use might be made of this Art."

Official investigation of all the petrified objects submitted to the Society fell to its curator of experiments, the multi-talented Robert Hooke. It was Hooke, the "mechanick genius," who took the philo-

sophical ponderings of other members and turned them into con-
crete, scientific experiments.

The air pump that Robert Boyle used for his famous experiments
on air, in which he proved that air is necessary for respiration, for
combustion, and for the transmission of sound, was designed and
built by Hooke. Hooke was, in fact, the only one who could get the
contraption to work.

Left to his own devices, Hooke was even more impressive. His
inventions included improved mechanisms for watches and pendu-
lum clocks, air gauges, wind gauges, and barometers. Building his
own telescope, he promptly discovered the Great Spot on Jupiter, cal-
culating the planet's rotation from its movements. With his microscope, he discovered plant cells. He even invented the term "cell" to
describe them.

Hooke was the right man to put on the fossil question. One of
the few Society members not born to wealth, Hooke, the son of a
clergyman, was raised on the Isle of Wight. As a boy, he was less interested in academic studies than in artwork and exploring the island's
fossil-ridden sea cliffs. Only after a failed attempt to train as an artist
(the smell of paint gave him a headache) did he turn to science.
Brilliant, always an outsider, Hooke never hesitated to take the unpopular side of a question. And he never forgot what he had seen in the
sea cliffs.

Not long after Hooke discovered plant cells, a piece of petrified
wood was brought to the Society and Hooke was instructed to view
it under his microscope. Cutting a thin slice, he saw it had the same
cellular structure as living wood.

Like other fossils, petrified wood was assumed to grow inside the
earth, stone and clay transmuted into wood. In 1637, Francisco Stel-

luti, an associate of Galileo, had even proved this scientifically. There were clay hills near Rome in which he found wood at all stages of the transformation, from solid rock to wood fresh enough to burn.

After seeing the cells in his petrified wood, Hooke thought Stelluti had it all backward. The stone did not turn to wood, the wood turned to stone. Buried in the clay, the wood had soaked in "mineral juices," which filled its pores, hardening it to stone.

Hooke soon became a vociferous critic of the idea that petrified wood, fossil seashells, and other fossils were "apish Tricks of Nature" generated within the earth.

Remembering the variety of fossil shells, bones, and teeth he had found in the sea cliffs as a boy, he asked why it was always these hard, decay-resistant parts that nature chose to sport with. "Why does it not imitate several other of its own Works? Why do we not dig out of mines everlasting vegetables, as grass for instance or roses of the same substance figure color and smell?"

To the mechanically minded Hooke, Kircher's mystic world of "plastic forces" and spontaneous generation was just so much nonsense—and he let everyone know it.

But he was decidedly in the minority. For most of Hooke's colleagues, Kircher's vision had struck a responsive chord. The mechanical philosophy endorsed by Descartes, and now their countryman Thomas Hobbes, was beginning to seem too materialistic, too close to atheism. Not everyone would vouch for Kircher's facts—his books always had a few howlers in them—but his "plastic forces" were a welcome balance to the cold, dead matter of the mechanical philosophy.

At Cambridge, a rising group of philosophers was advocating its own version of "plastic nature" and immaterial forces, attracting such devotees as a young mathematician named Isaac Newton. The shells in English rocks fit in beautifully with their friendlier view of nature. When Hooke charged that nature did nothing in vain, and that seashells generated in stone could serve no purpose, they had a ready reply: The shells were there to adorn the subterranean world, just as flowers adorned the world above.

Unfortunately, Hooke saw Steno's entry into the debate less as support for his position than as encroachment onto his turf. Hooke went so far as to accuse Henry Oldenburg, an admirer of Steno, but no friend of Hooke's, of sending word of Hooke's ideas to Steno, "by stealth," in order to undercut Hooke's career.

Hooke was almost as well known for his difficult personality as for his technical brilliance. According to one acquaintance, Hooke was "the most ill-natured and conceited man in the world." He was "despicable, being very crooked," said the editor of his posthumous works. The crookedness came from a growth deformity that prevented him from standing straight. He was vain and quick to take offense. At the theater one night, Hooke was outraged by a production of Thomas Shadwell's "The Virtuoso" in which a harebrained scientist, Nicholas Gimcrack, "weighs air" and spends all his money on microscopes and scientific gewgaws. " 'Tis below a Virtuoso, to trouble himself with Men and Manners. I study Insects," says Gimcrack. Hooke recognized himself. "Damned Doggs. People almost pointed," he wrote in his diary.

But his worst battles were with his fellow scientists. Hooke was notorious for his feuds with other members of the Society, accusing them of stealing his ideas and other crimes. According to Hooke, Isaac Newton used ideas pilfered from him as the basis for the theory of gravitation. The two became bitter enemies. After Hooke's death, his official portrait at the Royal Society mysteriously disappeared. The man in charge at the time was Isaac Newton.

When Oldenburg's translation of *De solido* came out in 1671, the battle lines were already drawn.

The first response was a letter to the *Philosophical Transactions* from a York physician named Martin Lister. A newly admitted member of the Society, Lister had met Steno in France while a medical student at Montpellier. The encounter had been friendly. Lister wrote home that he had seen a dissection by the famous Dane, "whom I found infinitely taking & aggreable in conversation & I observed in him very much of ye Galant & honest man as ye french say, as well as ye schollar . . ."

The tone of Lister's letter to the *Philosophical Transactions* was much different. Since returning from France he had, in addition to building his medical practice, earned a reputation as one of England's most competent naturalists. In fact, he was its foremost authority on seashells.

Lister's attack was withering. He did not doubt Steno that in some Mediterranean countries seashells might very well be found "promiscuously included in Rocks or Earth," particularly along the coasts.

"But for our *English* inland Quarries" he said, "which also abound with infinite number and great varieties of shells, I am apt to think, that there is no such matter as Petrifying of Shells in the business." The "Cockle-like stones" were stones and had always been stones. They had never been any part of an animal.

While "those persons, that think it not worth the while exactly and minutely to distinguish the several species of the things of nature" might disagree, he continued. "When they shall please to condescend to heedful and accurate descriptions, they will, I doubt not, be of that opinion, which an attentive view of these things led me into some years ago."

Lister had two main points of contention. First, he said, "there is no such thing as shell in these Resemblances of Shells." The material of fossil shells was completely unlike that of living ones. "It is certain that our English Quarry shells (to continue that abusive word) have no parts of a different Texture from the rock or quarry they are taken," he said, "that is, but that Iron-stone Cockles are all Iron stone; Lime or Marble, all Limestone or Marble; Sparre or Chrystalline-Shells, all Sparre, &c."

Lister knew his shells. But his grasp of fossilization was tenuous. Steno had explained at length in *De solido* how buried organic materials could be replaced by stone. As the son of a goldsmith, he knew how easy it was to take a mold of an object made from one material and then cast a replica out of something else. And just as he explained that mineral crystals could precipitate from fluids particle-by-particle, a shell or bone could be incrementally replaced the same way.

Lister's second argument was far more troubling. "Quarries of different stone yield us quite different sorts or species of shells" said Lister, "not only from one another . . . but, I dare boldly say, from anything in nature besides, that either the land, salt, or fresh water doth yield us." Despite their superficial resemblance to mollusk shells, fossils were "not cast in any *Animal mold*, whole species or race is yet to be found in being at this day."

This was indeed a question Steno had not addressed. The main shell-bearing strata in Tuscany are mostly of Pliocene age—that is, two to five million years old. Five million years is a long time on the scale of human evolution (the human species is only about a half million years old), but short on the scale of molluscan evolution. Most mollusk species from that time are still around today.

In England, the story was different. The fossil-bearing rocks in which Lister and others found their shells are mostly from the Jurassic and Carboniferous—as much as three *hundred* million years old. No species of mollusk from that time is extant today. *Whole families* common during that time are now completely extinct. England's climate was tropical and some of the fossil shells are gigantic.

Most distinctive, and most puzzling to seventeenth-century science, were fossil shells called "ammonites." These shells belonged to squidlike relatives of the chambered nautilus, which lives today in the depths of tropical seas. The spiral, chambered shells, often filled with crystals, sometimes as large as a wagon wheel, looked like nothing that had ever washed up on an English beach. Folk wisdom had it that they were petrified coiled snakes. Another theory was that they had crystallized from eddies in the subterranean vapors.

What could explain these anomalous forms?

There was, of course, the possibility that the unknown species, even the huge ones, still lived somewhere in the ocean's unexplored depths. But that raised the question of why it was the deep-water species, not the familiar shallow ones, that were fossilized. What kind of inundation would scoop shells up from the abyss, yet bypass the shallow waters?

Alternatively, perhaps the species in ancient seas were simply different from modern ones. This seemed most likely to Hooke, who toyed with the idea that the fossil forms might have evolved into present-day species. Like Darwin almost two hundred years later, Hooke saw the diversity of wild and domestic species as evidence of their adaptability. "We see what variety of Species, variety of Soils and Climate, and other Circumstantial Accidents do produce," he

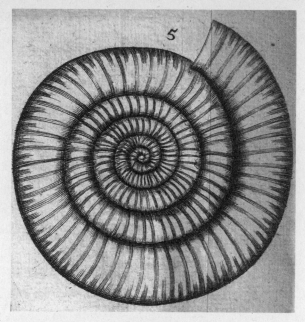

Fossil ammonite, measuring two feet across, illustrated by Martin
Lister in 1692. Martin Lister, *Historiae animalium angliae* (London,
1678). Courtesy of History of Science Collections, University of
Oklahoma Libraries.

said. God could easily have done the same thing with ammonites, al-
beit on a larger scale.

The suggestion was not completely out of line. Ever since the
discovery of the Americas, with their abundance of previously unknown animal species, Biblical literalists had been in a tight spot.
How could all these species have fit on Noah's Ark? Why were the
species on other continents different from European ones? A handy
solution was that by interbreeding and "accidental variations in process of time," Noah's original stock of animals had changed since the
flood. The non-European species were, naturally, degenerate versions
of the standard European forms.

But as an explanation for marine fossils, alteration of species was unacceptable to most of Hooke's contemporaries. The anomalous seashells in English rocks implied a total revamping of marine life. Why would God do that to what was supposed to have been an already perfect creation?

Linked to evolution was another possibility, dreadful to contemplate: extinction. Hooke—for whom evolution could fill in the holes left by extinction—was perfectly comfortable with the idea. But to others, such as the naturalist John Ray, the loss of even a single species was abhorrent, a "dismembering of the universe" that rendered it imperfect.

For Ray, who was to become England's greatest naturalist until Darwin, fossils were a genuine paradox. Like Lister, he had met and been on friendly terms with Steno in Montpellier. A blacksmith's son, he had trained for the ministry, only to be barred from preaching after he refused to take the oath of conformity imposed by King Charles. He found work as a tutor to the son of a wealthy gentleman, who encouraged his studies of natural history on the side. Ray took his charge on an extended educational tour of Europe. When they met Steno, they were on their way back to England, fresh from a fossil-collecting trip to Malta.

Ray read and praised *De solido*. From what he had seen in Malta, he agreed completely with Steno's arguments. He could hardly believe that nature could be "so wanton and toyish as to form such elegant figures without further end or design than her own pastime or diversion." The shells had to be real shells.

Except, they couldn't possibly be real shells—because if they were, the species they represented must be extinct. Ray believed that "divine providence is especially concerned to preserve and secure all the works of creation." Extinction was impossible.

The world's greatest scientific minds had turned their attentions to the question, yet fossils were confusing as ever. In *De solido,* Steno had promised that his full dissertation would resolve all lingering doubts.

Where was it?

HEAVEN AND EARTH

Truth shall spring out of the earth;
and righteousness shall look down from heaven.

—*KING JAMES BIBLE*, PSALM 85

U pon learning of Ferdinando's illness in the spring of 1670, Steno had rushed back to Italy. But by the time he got to Florence, the grand duke was already dead. Ferdinando's Cimento had been a paragon of the new scientific methods, but when he fell ill, struck by apoplexy, his caregivers fell back on traditional cures. They bled him profusely. They scorched his skin with red-hot cauterizing irons. They forced medicinal powders up his nose, and mashed live, partially disemboweled pigeons against his face. His final days were probably as undignified as they were uncomfortable.

The new grand duke was Ferdinando's twenty-eight-year-old son, Cosimo III. The boy had inherited few of his father's intellectual gifts. Cosimo's mother had vetoed Ferdinando's attempts to give their son an education in science and the new philosophy. She raised him in the old way, which meant in the strictest adherence to church orthodoxy. As a result, he was a young man "of singular piety," but also "melancholy to an extraordinary degree." Whereas Ferdinando had been "affable with everyone, as ready with a joke as a laugh," said an ambassador, "the Prince is never seen to smile."

Now on his father's throne, with his mother looking over his shoulder figuratively if not literally, Cosimo instituted a few changes. Raising taxes, he spent the money on redecorating the palace, holy relics (mostly of dubious origin) for his private chapel, and sundry projects to elevate the popular piety. He turned the Office of Public Decency loose on the city's sinners: Prostitutes were flogged and fornicators beheaded. May Day celebrations and other "pagan" rituals were chased from the streets, replaced by solemn processions of bloody flagellants and hellfire sermonizing.

Fearful of any departure from orthodoxy in his domain, Cosimo cracked down on the intellectual liberties that professors at the university in Pisa had enjoyed under his father. Tuscan students were forbidden to study anywhere else, lest they be contaminated by foreign ideas. In a move that must have put Ferdinando spinning in his grave, he banned all teaching of Galileo.

He showed a gentler side to the scientists under his own roof. In his own way, he tried to accommodate them. Prisoners condemned to be executed, of which there were plenty thanks to Cosimo's harsh new policies, could be strangled by hand rather than hung from the gallows, should his anatomists Steno or Redi prefer their cadavers with intact necks.

But the scientific spirit of the Cimento withered under Cosimo. The remaining academicians found alternative support or alternative vocations. Viviani now drew a pension from Louis XIV in France, to whom Cosimo was related by marriage. Magalotti gave up scientific research altogether to pursue a literary and diplomatic career.

Only Francesco Redi was able to continue more or less as before.

As Cosimo's personal physician, he had a secure position—the young grand duke was as terrified of death as he was worried about the salvation of soul. From his earlier studies of excrement and rotten meat, Redi had moved on to internal parasites. The steady stream of cadavers from Cosimo's gallows provided a wealth of material. He called his treatise on the subject "On Animals inside of Animals," an affectionate nod to his friend's unfinished dissertation on solids in solids.

This was the project on which Steno now struggled to focus himself. A converted heretic, earnestly pious, Steno held special favor in the eyes of Cosimo, who enthusiastically supported Steno's geological work. He set aside a small villa with a garden on the banks of the Arno where Steno could work undisturbed.

Steno had come back to Florence full of enthusiasm for completing the dissertation. Everything he had seen on his long journey had confirmed what he had written in *De solido*. In the mines of Hungary and Austria, he had seen that the veins of ore were no different from those in Italian mines: The veins came after the rocks, they were not from "the beginning of things." In the Alps, where the rock strata were fractured and highly contorted by the geologic forces that had raised the mountains, he found that even the most steeply tilted beds, upon examination, could still be seen to be layers of sediment originally laid down horizontally in water. Just as in Italy, the strata were a record of the earth's great changes.

But the work went slowly. As before, he always seemed on the verge of completing the dissertation—there were just a few more details to check, some more supporting evidence to gather. In his first two years back in Italy he traveled, visiting caves, mines, and other sites to gather yet more information. He also set to work organizing the ducal mineral collection, which he had greatly expanded by the minerals he had sent back during his travels. He fussed over his dissertation, which he was determined to write in Italian, as a tribute to his patrons and to Florence, which by now he was calling his "second home."

Ever since returning to Italy, Steno had been spending more and more of his time on theological studies in addition to his scientific work. Some of this was a continuation of the reading he had done for his conversion. But mainly it was a response to the furor his conversion had raised. For a well-known Protestant scientist such as Steno to convert was big news. And Steno had found in Amsterdam that many of his old friends did not take kindly to hearing it.

Stung by attacks on his conversion, Steno felt obliged to justify himself. With Cosimo's blessing, he devoted part of his time to writing responses to his critics. These documents, in the form of letters to Calvinist preacher Johannes Sylvius, give important clues to Steno's thinking as he drifted more and more toward the church. The main issue was the Bible. Steno had been raised to believe "read the Scriptures and you will know the truth," but now more than ever he doubted that that was enough. Which version of the Scriptures should he read? he asked Sylvius. Which was divinely inspired? And whose interpretation, among all the conflicting parties, should he accept?

The Bible was inspired by God, but was not his literal word, said Steno, only an approximation. The insistence on relying on the Bible alone for faith had led to divisiveness and hypocrisy.

Steno also wrote to Baruch Spinoza, with whom Steno had the opposite problem. Spinoza had just published his *Theological and Political Treatise*, in which he criticized organized religion. Spinoza's enemies claimed that his rejection of the Bible, and his focus on reason instead of faith, was no different from atheism.

The opposition to Spinoza's book was violent, even in tolerant Holland. Steno, who had known Spinoza there, was comparatively mild in his criticisms, but still he assailed Spinoza's system as a "dangerous" philosophy, "a religion of bodies not of souls." "You do not think there exists any certainty besides demonstrative certainty," he said, "and are ignorant of the certainty of faith which is above all demonstration."

Spinoza's intellectualized approach to religion, Steno charged, was a realistic option for "the learned alone" and those "free from the vicissitude of business." Only the Catholic Church offered "unity of the beliefs and actions" and salvation without distinction to all persons of every sex, age, and condition. There is no record of Spinoza's response; in fact, he may never have seen the letter.

From these writings, it is plain that Steno struggled to find a balance between the overly literal interpretation of the Bible of his Protestant background and the overly rationalistic views of nature of some of his scientific colleagues.

Steno had been busy trying to convert others, and he did win some. Among them was Spinoza's close associate Albert Burgh, which resulted in hard feelings between Burgh and Spinoza. Steno's newfound zeal for religion also caused friction with some of his old friends. "Steno is all too partial, and he only thinks of making someone catholic," Jan Swammerdam wrote to Thévenot. "I wish he were still like he was when he sought God in the bible of nature."

Steno's geological and theological studies were interrupted two years after his return to Florence by another summons to Denmark. This time the offer came with a guarantee of religious freedom. Steno reluctantly accepted the offer. He was still a Danish subject and felt he had no choice but to comply with the summons.

It was in Copenhagen in the winter of 1673 that Steno made his final public appearance as a professional scientist. Dedicating the re-vamped anatomical theater, he gave an opening statement on the importance of anatomical research. But he now viewed science as simply a way station to more spiritual questions, uttering his famous aphorism:

> Beautiful is what we see.
> More beautiful is what we understand.
> Most beautiful is what we do not comprehend.

He lasted two unhappy years in Copenhagen. He missed Florence, and missed his geological studies, writing to Cosimo that in flat, sandy Denmark he had to limit his investigations to trenches excavated during building construction. Plans for a fossil-hunting trip to Sweden fell through. His friends Bartholin and Borch were uninterested in his geological theories.

In the end, finding less support for his science and less freedom of religion than he had been promised, Steno petitioned to the king to be released from his obligations and be allowed to return to Florence.

Steno's career as a scientist was rapidly drawing to a close. Back in Florence he took the position of tutor to Cosimo's unruly eleven-year-old son, prince Ferdinando III, a job for which he soon learned he was completely unsuited. Rebelling against his father's excessive piety, intelligent and artistic, but completely uninterested in his studies, the prince resisted all of Steno's earnest attempts to reform and educate him.

But Steno was now convinced he had a higher calling, anyway. Protestant scientists visiting Florence could always be sure that Steno would make a pitch—in his gentle, soft-spoken way—that they, too, come back to the Church. In Denmark he had opened an anatomical lecture with the statement: "The true purpose of anatomy is to lift the beholder of the body's wondrous work of art up to an admiration of the dignity of the soul and thence to the knowledge and love of the Creator." For Steno, anatomy had now fulfilled its purpose, and he prepared for a new career. He would become a priest.

In view of his reputation as a scholar, the Church waived the customary theological examination and in April 1675, barely four months after his return from Denmark, he was ordained, saying his first mass at Easter. He took an immediate vow of poverty.

Steno's new position in the Church did not preclude scientific research. Far from it. Many scientific researchers were members of the clergy. Probably many of Steno's friends in Florence expected him to continue as before. But Steno was not one for half-measures. He told Athanasius Kircher, with whom he had become friendly despite their differences on fossils, that he was giving up science as a sacrifice to God.

But for years he had been talking with friends about the limits of scientific research, how inadequate it was as a means for reaching "divine certainty" regarding issues of the soul. He had also soured on the competitiveness of science; the open-minded investigation of nature he hoped it would be had too often degenerated into petty rivalries and scrambles for fame.

After becoming a priest Steno remained on Cosimo's payroll, serving as the grand duke's confessor, and trying, without success, to persuade him to reduce the crushing burden of taxes imposed on the city's population. Devout and severe as Steno was regarding his own faith, he worried about imposing overly strict religious codes on the general populace. They tended, he said, to destroy a lot of good grain along with the chaff.

He also was still tutor to the prince, weaving sermons on Christian humility into the boy's lessons on natural science. One lecture to his pupil on gems and crystals, which still survives in manuscript, cautions against the vanity and sense of self-importance that results from wearing such baubles. Instead, the beauty of the stones should inspire reflection on the agonies of the laborers who mined them, and the upheaval of the earth that created them. Steno's geological homilies evidently had no lasting effect. The Grand Prince Ferdinando went on to live the high life with the same gusto as any other wealthy nobleman, eventually dying of syphilis before he could rise to the throne.

Steno had not, however, completely forgotten about *De solido* and his unfinished geological research. Three Danish students, Thomas Bartholin's two sons and a nephew, had followed him from Copenhagen to study anatomy in Florence. Steno not only put them to work in the dissecting laboratory, but also sent them into the hills to collect fossils. Bartholin's nephew, Holger Jacobaeus, seemed particularly interested in geology and, not coincidentally, in Steno's Catholic faith as well.

For Steno's first two years as a priest he held no clerical office in the Church. He intended to devote himself to entirely pastoral work, rather than theology or advancing in the hierarchy. In the spring of 1677, however, that abruptly changed when he was summoned to Rome and made a bishop. No one was more surprised than Steno, and it was, in fact, more administrative responsibility than he wanted. But the pope saw an acute need for a bishop to minister to the small remnant population of Catholics in Germany. Cosimo, who had long harbored grandiose visions of reconverting the heathens across the Alps, volunteered to bankroll Steno's mission. Scandinavia, which also still had a few stray Catholics, was added to his area of responsibility.

Steno had made the journey to Rome for his consecration on foot, living on alms along the way. Six months later, Bishop Steno climbed aboard a mail coach that would take him through the Alps, where he could look up at the high peaks and convoluted strata of the mountains one last time.

In Germany, he would devote his energy to converting the lost souls of Lutherans back to the mother Church. More important, as a missionary for his new science, he would make one final convert.

14

VAIN SPECULATION

One cannot conceive anything so strange and so implausible that it has not already been said by one philosopher or another.
—RENÉ DESCARTES, *DISCOURSE ON METHOD* (1637)

While Steno embarked on his new life, choosing to save souls rather than study seashells, the fossil debate continued. In 1678, as he began his missionary work in Germany, an Amsterdam publisher reprinted *De solido*.

Martin Lister, England's leading student of seashells, continued to assert in his abrasive way that only dilettantes not bothering to make "heedful and accurate" descriptions could claim that fossil shells were the genuine article—a close look revealed innumerable differences between living mollusk species and their stony mimics. "Quarry shells," as Lister called them, were purely mineral features, the work of "shooting salts" inside rocks; a point he tried to prove by growing seashells in beakers of mineral solutions and body juices of mollusks.

Lister's experiments failed, but his ideas remained popular just the same. There was no need to confirm them in the laboratory when nature herself offered abundant proof. Fossil shells of different sizes and in varying states of dissolution were interpreted to be at different stages of formation inside the rocks. Such things, wrote one Royal Society member to Robert Boyle, "which we have seene with our eyes halfe generated, and in transitu towards Generation," seemed "an unanswerable argument for their never having been the spoils of Animals."

As to the contrary opinion, said a letter to the *Philosophical Transactions* published in 1676, the writer agreed with Lister that "it seems not to be grounded on practical knowledge."

After a thorough study of the figured stones in the mines of Somerset shire, he came to the conclusion that mineral growth was the only logical explanation. And why not? There are many gradations in nature, he said. And if minerals could grow in the sea—he gave coral as an example—there was no good reason "why Shells might not as well be produc'd in Mines."

When Lister published his magnum opus on seashells in 1678, which amplified his earlier arguments against Steno's explanation of fossils, he was largely preaching to the converted. Lister was soon joined by the world's other preeminent expert on shells.

Filippo Buonanni, a student and disciple of Athanasius Kircher and among the last of the Renaissance Aristotelians, knew too much about seashells to ignore the differences between living and fossil species. He had collected shells since his school days in the port city of Ancona, where he had amassed his own museum. Now a Jesuit priest in Rome, Buonanni published *Recreation for the Eye and Mind in the Study of Shells,* the first illustrated guide to seashells ever, as a tribute to his beloved mollusks. Spontaneous generation was an incontrovertible fact, he believed, regardless of what Redi claimed to have proved with his flies and veils. Aristotle had taught this was the invariable mode of reproduction for shellfish. The seashells in the hills could therefore have no other origin.

Buonanni doubted that Noah's flood could have swept so many shells so high into the mountains in such a short time, and, ignoring Aristotle's doctrine of natural inundations, he saw no other way to put marine remains on dry land. Steno, he said, was sadly in error.

The "Wise Aristotelian," as Buonanni was called, was provoked not only by Steno, but by another book published a couple of years after *De solido.*

Written in lively and polemical Italian, *Vain Speculation Undeceived by Sense* was the work of Agostino Scilla. Unlike the experts of conchology, he presented himself as a "simple man," "nude of learning." The book was devoted entirely to the subject of marine fossils, and he wished only to make a simple point: What looked like seashells in the hills *really were seashells.*

No amount of "high philosophizing" by scholars could change this obvious and straightforward fact. He knew that philosophers scorned appearances, preferring to "take flight with the intellect to distant and spacious fields of the possible," but he knew seashells when he saw them.

"It seems to me impossible to arrive at any sort of knowledge of the truth," he said, "if I abandon the path my eyes show me." If sense "has tricked me," he asked, "to whom could I turn?"

Scilla was not so simple as he let on. A successful landscape painter and antiquarian, he was well-connected in Sicilian and literary circles under the name *lo Scolorito,* "The Faded One." Against "the broken, dismembered and, to speak clearly, dreamed-up," theories of the modern philosophers he was merciless.

But neither Scilla, nor anyone else, had a good answer for the question of how the sea could have risen so high and overtopped the mountains. Pledging "to live and to die under the dictates of the Holy Roman Church," he could only point out that even the church

Frontispiece of Scilla's book. Agostino Scilla, *Corporibus Marinis* (Rome, 1747). Courtesy of History of Science Collections, University of Oklahoma Libraries.

fathers and the greatest of theologians had been of many opinions regarding "the manner God chose for drowning this world."

As for himself, he did not know whether the shells were from Noah's flood or other "special floods." "Neither do I know whether this beast of a World . . . at some particular time, tired of lying on one side, might have turned over and exposed its other side, which had been under the water and was full of so many bits of the refuse of the sea, to the rays of the Sun." Anything was possible, but what *actually* happened so long ago in the past, "I do not know, neither do I know the way to find out. Nor do I care."

There was, of course, a way to find out. Steno had laid out the basics in *De solido*. But so long as authorities such as Lister and Buonanni denied the biological reality of fossil seashells, and so long as the theories of plastic nature held sway in England and Rome, there was no reason for anyone to try. Steno's principle of superposition, sensible as it was, was irrelevant if strata weren't beds of sediment. And if the fossil seashells weren't seashells, there was no reason to believe that they were.

What's more, the kinds of geologic upheavals Steno's theories required were becoming less and less appealing for even the most scientific minds. Even the tumult of the Flood, which no one doubted had actually occurred, was becoming tamer. While a hundred years earlier Martin Luther had described it as an all-consuming cataclysm, rending the earth and piling even mountains atop one another, now the waters were assumed to have been deep but placid.

The evidence was in Scripture. After the waters had receded, Noah released a dove, which returned with an olive leaf in its beak. How could the Flood overturn mountains if it couldn't even uproot an olive tree?

The evidence came from science, too. The estimable Robert Boyle published a pamphlet showing that the deep waters at the bottom of the sea were tranquil, undisturbed by even the most violent storms. Though Boyle himself was a great admirer of *De solido* and was inclined to agree with Steno's interpretation of fossils, his deep-sea pamphlet was trotted out by critic after critic well into the eighteenth century.

Most important, though, was that the idea of the angry, punitive God espoused by Luther, John Calvin, and others a century before was falling out of fashion. The era of plagues and religious warfare seemed to be drawing to a close. In England, especially, prospects were looking up: the decades of civil war and strife between factions of the Anglican church had been brought to an end with the restoration of King Charles to the throne.

God was showing his benevolent side again, it seemed. Nature was a symbol of Divine wisdom, not human sin. It revealed the logic of God the Geometer and the ingenuity of God the Engineer. No one believed this more deeply than the scientists of the Royal Society.

In answer to the old belief that the world was in a state of decay, as evidenced by the continual erosion of the landscape, the young Isaac Newton suggested that the surface of the earth was constantly being renewed from the heavens by a constant rain of ethereal matter. (The downward pressure caused by this "food of the Sunn & Planets" also explained gravity.)

In this new climate, Kircher's everlasting mountains lovingly shaped by God with his own hands, mountains designed to be the world's superstructure, appealed to the hearts and minds of Christian scientists. Steno's ragged mountains and collapsing strata did not.

"The Under-ground World is a well-framed House," wrote the editor of an English translation of Kircher's book, "with distinct Rooms, Cellars, and Store-houses, by great Art and Wisdom fitted together; and not as many think, a confused and jumbled heap or Chaos of things, as it were, of Stones, Bricks, Wood, and other Materials, as the rubbish of a decayed House, or an House not yet made."

The world was created not just for human utilitarian needs,

however. It was specifically intended to be a stage for the Christian drama.

"The universal mechanism of the world was foreseen and foreordained from eternity to this end," wrote Kircher, "it came into existence not just for its own sake, but so that it might be of service to the earth, which is the beginning and the end of the entire universe, and which must work together with all the forces of the heavens, without which it could not have been preserved, for the salvation of the human race."

The indecisive John Ray, keeping one eye on Steno's troublesome and possibly extinct fossils, chimed in with his own praise for the wisdom of the earth's design. He wrote a book, *The Wisdom of God Manifested in the Works of Creation*. Mountains, which for so long had been reviled as blemishes on the face of the earth, were admirable things, said Ray. They provided the sources of the springs and rivers necessary for life, and served as habitats for useful plants and animals. They were cleverly designed for the "convenient digging up of Metals and Minerals." God had even had the foresight to lay them out in such a way to create "Boundaries and Defences to the Territories of Kingdoms and Common-wealths."

The earth, formerly unworthy of a history, was now too good to have one.

And while many still believed that its prophesied six-thousand-year lifespan was nearly over, the attitude now was not dread but eager anticipation. Human history was about to reach its climax. Science played a role in this by bringing the human mind ever closer to perfection. When Newton published his theory of gravitation, one of his disciples described it in Biblical terms: "an eminent prelude and preparation to those happy times of the restitution of all things, which God has spoken of . . . since the world began, Acts iii 21."

There was no place for speculations about geologic upheaval in the emerging new scientific ideal. There was, however, a place for immaterial forces such as magnetism and, as Newton would prove, gravity. A few years earlier a letter in *Philosophical Transactions* had seemed to confirm the reality of plastic forces as well.

In the letter, a physician reported that he had extracted the small, perfectly formed shell of a marine snail from a woman's kidney. It appeared to be a decisive blow against the organic theory of fossils because, as mathematician John Wallis wrote to astronomer Edmond Halley, the shell was "more likely to have been formed there, than that this Kidney had been Sea."

Though Robert Hooke pointed out that even if the woman's kidney had never been immersed in the sea, the surgeon's sponge surely *had,* and was therefore a more likely source of the shell than the kidney, Wallis summed up the attitude of most of his colleagues. "Sea where now is sea, and Land where now is land," he said. Great changes on the earth are "contrary to the History of all Ages."

John Ray had to agree with Wallis on the stability of geography. Yet, as England's most highly esteemed naturalist, he could not so glibly dismiss the evidence of fossil shells. Their biological appearance was undeniable; Lister's theory of "shooting salts" was vague and unpersuasive, but the implications were shocking. Foreshadowing debates to come, he reasoned that if the shells in the mountains were real, "the world is a great deal older as imagined or believed, there being an incredible space of time required to work such changes as raising all the mountains."

15

THE BISHOP

He putteth forth his hand upon the rock; he overturneth the mountains by the roots.

He cutteth out rivers among rocks; and his eye seeth every precious thing.

He bindeth the floods from overflowing; and the thing that is hid bringeth he forth to light.

But where shall wisdom be found? Where is the place of understanding?

—JOB 28:9–12

Steno spent his career as a bishop, and the last years of his life, *in partibus infidelium*, in the land of the infidels. In Hannover, which would serve as his first base of operations, his duties were to minister to the few Catholics in northern Europe who had survived the Thirty Years' War, and to bring back into the fold as many Protestant heretics as possible. The clergy was spread thin: Steno was responsible for northern and western Germany, Denmark, and Norway.

He was a titular bishop, which meant he had no see; that is, he had no cathedral, no seat of ecclesiastical authority. He was nominally bishop of Titiopolis, an ancient see somewhere in Asia Minor, which had long since been lost to the Turks. Its actual location was unknown. There was nothing unusual about this arrangement. Bishops

sent into hostile Protestant territory were often put in charge of
nonexistent sees. Steno's predecessor in Hannover was the bishop of
Morocco.

Given the local hostility to Catholics and the still-difficult economic situation even thirty years after the war, Germany was considered a hardship post. Hannover was one of more than three hundred
petty German states that formed the tattered remains of what was
called the Holy Roman Empire (though Voltaire later pointed out
that it was none of the three). It was one of the few German states
that tolerated Catholics at all.

Like Steno, the Duke Johann Friedrich was a Catholic convert.
They had met before when Steno had visited Hannover during earlier travels, and Johann Friedrich had specifically requested that
Rome send Steno to be his bishop. Equally eager to have Steno come
to the city was the duke's librarian, who pined for intellectual com-
panionship in the isolated and somewhat provincial city. He was the
brilliant philosopher and mathematician Gottfried Leibniz.

Remembered mainly for his philosophy of "preestablished har-
mony" on one hand and his bitter feud with Isaac Newton over the
invention of calculus on the other, Leibniz was, according to one
modern writer, "the last universal genius." His famous claim that this
is the "best of all possible worlds" was later ridiculed by Voltaire in
Candide, but for Leibniz things *did* always seem to turn out for the
better.

He liked to tell the story of the secret society of alchemists he had
once applied to join. To prove his knowledge of their mysterious art,
he put in his letter random phrases copied from books. Although he
understood not a word of his own letter, they not only admitted him
into their circle but asked him to serve as secretary, even offering him
a pension.

"It is so rare," said the duchess of Orleans after meeting Leibniz, for an intellectual "to be smartly dressed, and not to smell, and to understand jokes."

Leibniz had been in Hannover for two years, coming from Paris to take the librarian job after his previous patron had died and Leibniz had run out of money. As was his practice, he kept himself busy by doing everything besides what he was paid to do. He worked on his mathematical and philosophical theories, tended to his voluminous correspondence, and peppered the duke with ideas for political projects, business ventures, and inventions (all to be financed by the duke, of course). There was no time for cataloging books.

Leibniz had an abiding interest in theology, and was eager to share ideas with the new bishop. He proposed several elaborate schemes for reuniting the Protestant and Catholic churches, none of which seemed realistic to Steno, who met Leibniz's optimism with his own pessimism. Besides, as was his wont, Steno was now wholly committed to Catholicism, giving Protestant heresies no quarter, even if he maintained friendly relations with Leibniz and other Protestants.

Most of what is known about their relationship comes from Leibniz's notes and letters. He dramatized one of their theological conversations in an unpublished dialog in which Théophile (Leibniz) and Poliandre (Steno) debate the nature of God and free will, and Leibniz's belief in the "best of all possible worlds." Théophile naturally comes out on top; Poliandre inevitably is forced to concede his point.

In reality, there was no such meeting of minds on theology. And for his part, Leibniz deplored Steno's decision to leave science for religion. "From being a great physicist he became a mediocre theologian," said Leibniz. Among Steno's scientific works, he particularly admired *De solido* and urged Steno to enlarge it. Leibniz had a longstanding interest in fossils, and until he had encountered Steno's ideas he had been in Kircher's camp. Now he was an enthusiastic convert.

Better than others, perhaps even better than Steno, the expansive Leibniz recognized the enormous potential of Steno's new science of the earth's strata. Applying it on a larger scale, one could, he said, "draw from it conclusions regarding the origin of the human race, the universal Flood, and other great truths . . . in the Holy Scriptures."

Steno most likely agreed, but someone else would have to do it. Except for the occasional dissection to demonstrate the errors of Cartesian anatomy, he considered himself through with science. Religion was now his entire life. He was so intent on devoting himself utterly to the church that, to Leibniz's consternation, Steno sometimes refused even to talk about science.

Unfortunately, when Johann Friedrich died in December 1679 and his Lutheran brother, Ernst August, was installed as the new duke, even these unsatisfactory conversations had to end. As most Catholics were forced out of the city, Hannover no longer needed a bishop. After barely two years in residence, Steno had to leave.

The last years of Steno's life were by far the most difficult. If his idealistic visions of science had been dashed by the realities of academic politics, his dreams of a purely spiritual life in the Church suffered even worse. Frustrated by bureaucracy and corruption in the Church and indifference among the laity, he sought religious solace by taking his vows of poverty and self-denial to ever-increasing extremes. He had no taste for administration or for ecclesiastical politics, and he repeatedly asked Rome to release him from his position as a bishop. He wanted to work as an "ordinary priest."

From Hannover he fled to Münster, where as auxiliary bishop he tried to reform the lax ways of the local clergy, with little effect. Steno's precise mind, which served him so admirably as a scientist, now fixed itself on the canons of the Church as the only path to salvation. But few mortals could live up to his exacting standards. He especially made enemies among the wealthy parishioners and upper ranks of the clergy by his advocacy for the poor and his criticisms of the Church's questionable financial practices. When the head bishop died and the new bishop bought his way into office, Steno rebelled, refusing to conduct the mass for the occasion. Again he had to flee, this time driven by angry Catholics.

He next settled in Hamburg, where his lifestyle became more and more ascetic. " I found him there without a house, without a servant, devoid of all life's comforts, lean, pale and emaciated," wrote Johannes Rose, a friend and one of Steno's German converts. According to Rose, Steno sold his bishop's ring and silver crucifix and gave away the money. At night he slept sitting in a chair or in a bed of straw on the floor. When he wasn't praying and meditating, and when he wasn't too weak from his constant fasting, he tended to his pastoral duties on foot "dressed like a pauper."

Though Rose and other admirers saw Steno as a living saint, many in Hamburg were offended by his extreme self-mortification. Fellow priests viewed his ragged attire as beneath the dignity and propriety of a bishop. Hamburg's six hundred or so Catholic parishioners were, if anything, more hostile than those in Münster. They threatened to cut off his nose and ears and run him out of town. Some threatened to kill him.

Isolated, his confidence shattered, Steno wrote to a friend in Florence that he was "living as a dead man who feels nothing." His reforms had failed. He was terrified over the fate of his soul. In Hamburg, he found lodgings with an old friend, Theodor Kerckring, an anatomist and fellow student from his Leiden days. Some of Steno's interest in science revived, if briefly: he dissected a human heart and even made notes for an essay on the nervous system. More than anything, however, he wanted to go back to Florence, where he had been happiest.

For years the Church had refused to let him step down as a bishop or grant him leave from his episcopal duties. In 1685, it finally relented, promising him a sabbatical. He quickly wrote to Cosimo and made arrangements to return to Florence for a period of "spiritual rejuvenation."

But once again, for one last time, his plans were thwarted, and the trajectory of his life was unexpectedly diverted. Just days before he was to leave Hamburg for Florence, he was asked to postpone his trip for a few weeks to help set up a church in the medieval city of Schwerin. On arriving, however, he found the resident missionary priest too ill to work, and the Lutheran authorities uncooperative. The "few weeks" stretched into months.

Over the years Steno had progressively destroyed his health through his rigorous and self-imposed lifestyle. He suffered from a chronic and sometimes incapacitating colic. A portrait painted that year in Schwerin shows a weary and emaciated man.

On November 21, 1686, he was suddenly gripped by an intense abdominal pain. He continued his normal routine in spite of the pain, but two days later he collapsed and was carried to his bed, his belly "swollen and taut like a drum." He asked for pen and paper to write his will and some final letters. Writing to Kerckring in Hamburg, he gave a clinical self-diagnosis:

> To my usual ailment, colic, it seems now that the stone has been added. Last night I had the most terrible pains in the os sacrum. After an enema, they have shifted to below the os pubis, and from this morning it seems as if they are increasing, as if an inflamation is forming there. Not a drop of urine comes. I believe that the stone has

Bishop Steno, from a portrait painted in Schwerin around the time of his death.
Courtesy of the Institute of Medical Anatomy, University of Copenhagen.

embedded itself in a fold in the bladder, and that there, besides causing
pains, it is causing inflammation of the mucous membrane of the
bladder and will be the cause of my death.

The swelling of his abdomen became so severe he worried it might rup-
ture, and he was soon having difficulty breathing. Coincidentally, the
sickly resident priest had died just days before, so a Jesuit missionary
from a nearby town had to be summoned to deliver last rites. When the

priest still had not arrived by the following morning, Steno made his last confession of sins to his assembled household, and asked them to read prayers for the sick. After a time, he faintly told them to switch to a prayer for the dead.

Shortly before seven o'clock on the morning of November 25, 1686, Steno died. He was forty-eight years old.

At the time of his death, Steno had almost no worldly possessions aside from books and religious paraphernalia. According to Johannes Rose's inventory, Steno's clothing and personal furnishings consisted of "a wretched black garment, an old tunic, his old cloak, two sack-cloth shirts, some small worn handkerchiefs which he also wore as cravats, and a nightcap." The funeral was delayed for nearly two weeks for lack of proper clothing to dress the corpse.

In keeping with his instructions, Steno's funeral was "mean and poor." But many of Steno's friends were convinced of his saintliness. Johannes Rose and others in Germany and Florence immediately wrote testimonials and privately gathered documentation in hopes of having him canonized, but there were no official Church proceedings.

Steno did ultimately make it back to Florence. When Cosimo learned that Steno had died, he sent money at once to have the remains shipped to Florence, where they would be interred with Medici family members in the San Lorenzo Basilica. In May 1687, six months after Steno's death, the corpse was loaded onto the ship *Saint Bernhard,* bound for Italy. Because superstitious captains generally refused to allow dead bodies aboard their vessels, Steno made his final journey in a casket disguised as a crate of books.

16

WORLD-MAKERS

We should like Nature to go no further; we should like it to be finite, like our mind; but this is to ignore the greatness and majesty of the Author of things.
—GOTTFRIED WILHEM LEIBNIZ,
LETTER TO S. CLARKE (1715)

When news of Steno's death reached Hannover, Leibniz immediately began making inquiries about Steno's papers, hoping that among them he might find the uncompleted geological dissertation.

Over the six years since Steno had left Hannover, geology had become one of Leibniz's many obsessions. He was inspired not only by his conversations with Steno, but his own observations in the famous mining district in the Harz Mountains to the south of Hannover. He had an ambitious plan to write a complete history of the world, from its creation to the birth of his employer, the new duke, Ernst August.

He had first persuaded Ernst August to finance an ill-conceived scheme to build a system of windmills to pump water from the mines, whose operations were often curtailed by flooding. The specially designed windmills were Leibniz's own invention, and they would have worked had he built them in a place where there was wind. But he didn't, and the project was a failure. The duke was angry over the wasted money and the miners over their wasted time, but Leibniz was delighted at the opportunity to make a firsthand study of the Harz rock strata.

Afterward, the duke, hoping to put Leibniz's energies to better use, set him on an entirely different sort of task. Leibniz would write a history of the duke's family, the House of Brunswick. Such a history would (the duke hoped) shore up his political claims within the Holy Roman Empire by establishing the family's deep roots.

Leibniz leaped at the project and immediately began researching a history far more grandiose than even the duke had envisioned. Picking up where Steno had left off, he would begin the family history with the laying down of the earth's first rock strata. To the duke it probably seemed unnecessary to trace his family that far back, but Leibniz was Leibniz. And what he had seen in the Harz mines had convinced him that, contrary to Steno, it was Germany, not Tuscany, whose rocks were most worthy of study.

Searching for Steno's unpublished manuscript, he first went to Hamburg, where he found Kerckring, who knew nothing about any geological documents. All he could tell Leibniz was that everything was gone, shipped to Florence with the corpse. Not missing a beat, Leibniz immediately arranged an extended research-gathering trip through Europe, visiting genealogical archives and geologizing along the way, which brought him to Florence. But there, too, the geological papers were nowhere to be found.

Not until years later, when he was settled back in Germany, did Leibniz learn that when Steno had become a bishop, he had entrusted his geological manuscripts to one of the Danish students visiting him in Florence, Thomas Bartholin's nephew Holger Jacobaeus.

Steno may even have expected Jacobaeus to publish the long-awaited dissertation. But around the time Bishop Steno set out for Germany, Jacobaeus also made a hasty exit from Florence. It seems the Bartholin family was worried that Jacobaeus, under the influence of Steno, was developing Catholic sympathies and might even convert. To prevent that disaster, he was ordered home to Lutheran safety, leaving most of his possessions to be shipped afterward. Much of it never arrived, and that was the last anyone had heard of the manuscript.

Jacobaeus had come home to a career as a physician and academic in Copenhagen. If Leibniz ever reached him (an existing letter shows he tried), he was no help. Nor did he ever publish a word on geology, of Steno's or his own.

Aside from Leibniz's search for the lost geological papers, during the years immediately following Steno's death, interest in his theories of fossils and strata was at a low ebb. The consensus was that the organic theory of fossils was an idea that had come and gone. Even Steno's old mentor, Ole Borch, seems to have dismissed it. In 1687, he published a short dissertation "On the Generation of Stones in the Macro and Microcosm," solidly in the old hermetic tradition. Nowhere does he even mention Steno or *De solido*.

While Leibniz labored to forge a link between geology and genealogy in Germany, events were unfolding in England that would soon bring Steno, or at least his science, back to center stage.

It started back in 1680 with a controversial book published by an Anglican priest named Thomas Burnet, the king's chaplain. Burnet was inspired to write his *Sacred Theory of the Earth* by the mountains he saw during a trip through the Alps. Horrifying to behold, they seemed to him nothing but wild "indigested heaps of Stones and

Earth." He decided he would not rest until he could explain how "that confusion came into Nature."

But unlike Steno, Burnet did not collect fossils or examine rock strata to find his answers. After his initial mountain experience, Burnet avoided any more encounters with wild heaps of stones. Basing his theory on "Scripture, Reason, and Ancient Tradition," he philosophized indoors until at last a "new Light still broke in, both from the Holy Scriptures, and the Monuments of the Ancients."

The account he came up with married old-line pessimistic theology to newfangled Cartesian physics. Described in elegant literary style, the egg-shaped world of his theory was created without mountains or seas or any other blemishes on its perfectly smooth surface. At the time of the Flood the egg cracked open, unleashing the waters of the deep. All the planet's irregularities, especially its ugly, jagged mountains, date from this "first great revolution of Nature." So too does the earth's tilted axis: formerly the earth was upright and in a state of perpetual springtime.

In Burnet's view, the earth today was a pathetic shadow of its former self: a "World lying in its rubbish." It was, as he called it with open contempt, just a "dirty little Planet."

Burnet said he hoped that in addition to putting the sordid earth in its place, his book would "silence the Cavils of Atheists" by reconciling science and scripture. But he did this by keeping scientific facts to a minimum and by ignoring the literal text of Genesis when it didn't fit his physical scenario. From both sides came the predictable howls.

The *Sacred Theory,* for all its scientific pretensions, was "no more or better than a meer chimera or Romance," said John Ray the naturalist. Philosopher John Locke claimed he could not reconcile the theory "either to philosophy, scripture, or itself."

More particularly, it ran against the newly sanguine Protestant theology, held dear by Anglican divines and Puritan scientists alike, that the world was not a "great Ruine" irreparably defiled by human sin, but a wondrous creation that God expressly designed for the convenience and edification of His favorite species.

And, far from silencing the irreligious, Burnet's loose reading of Genesis, and his mingling of physics with matters of faith seemed more likely to encourage them. "That way of philosophizing all from Natural Causes I fear will turn the whole World into Scoffers" worried one churchman.

Whether or not Burnet's book was responsible for turning many people into "scoffers" is hard to say. What it did do, without question, is turn many into geological theorists.

The *Sacred Theory of the Earth* spawned a slew of imitators and counter-theories, much to the alarm of religious conservatives, who saw the specter of atheism and freethinking behind each new speculation. But to others, this surge of grandiose philosophizing inspired ridicule. In their amusement, London's literary wits dubbed the theorists "World-makers." Never especially friendly to science, the satirist Jonathan Swift and his cronies mocked every attempt "to reform the Architecture of the World; and make the Creation look a little more Mathematical; to discover the Globe of Earth to be only a large Work in a kind of Pastry."

A ballad popular in the London coffee shops had the world-makers singing:

> That all the books of Moses
> Were nothing but supposes;
> That he deserv'd rebuke, Sir,
> Who wrote the Pentateuch, Sir,
> 'Twas nothing but a sham.

Many scientists, such as Isaac Newton, who had recently achieved godlike status after laying down the laws of physics in 1687's *Principia,* also looked down on the World-makers. Like Kircher before him, Newton maintained that the vast machine of the world was created in its present perfect state by the direct action of God; it was "unphilosophical" to think otherwise.

But even the great Newton couldn't resist joining the fad and indulging in a little world-making of his own. Writing to Burnet, he speculated that some of the earth's topography might have resulted from the earth's initial creation, rather than Noah's flood. "Milk is as uniform a liquor as the chaos was," he explained. " If beer be poured into it and the mixture let stand till it be dry, the surface of the curdled substance will appear as rugged and mountainous as the earth in any place."

Other theorists, such as Newton's disciple William Whiston, invoked not curdled milk but close encounters with comets and similar catastrophes to explain the features of the land. Many held fast to Kircher's vision of the earth's immutable frame, and denied anything on its face had ever changed. The microscopist Antonie van Leeuwenhoek, after examining sand grains under his lens, marveled that their individual shapes had lasted since the first days of Creation.

What was missing from Burnet's theory and many of its kin, however, was an explanation for the things that had sparked curiosity about the earth's mutations in the first place—fossils.

Burnet did not mention them at all. In fact, according to his theory there should be no such thing as marine fossils in mountains. Formed by chunks of primordial crust uplifted during the Flood, mountains could not possibly contain seashells because up until that time there had been no seas anywhere.

Steeped in the teachings of the platonists at Cambridge, who were keeping alive the belief in nature's plastic powers, Burnet was untroubled by the fossil question. How the earth managed to generate oddly shaped stones within its bowels was irrelevant to his big story.

But to John Ray, who after twenty years still wavered between Steno's and Lister's interpretations, it was a glaring omission. Fossil seashells were enigmatic as ever, turning up not only on mountains, but in the deepest of mines. Ray and other naturalists were finding that many of the fossils unearthed in England and points north gave every appearance of being tropical species. Could elephants have lived in Oxford? Hippos in Kent? Corals and crocodiles in Germany?

Even more than Ray, Robert Hooke was eager to reopen the question of fossils, especially fossil seashells. Hooke shared none of Ray's qualms about fossils and extinction, and it still rankled him that his earlier arguments had not been taken seriously by his colleagues. The biggest stumbling block to accepting fossil seashells as genuine, namely, how they and their enclosing rocks could be raised above the sea, he now felt he could explain by a complicated new theory he had developed, involving a combination of earthquakes and shifts of the earth's rotational axis.

But rather than publish his ideas, as the other geological theorists were doing, Hooke dusted off his old notes and declaimed them in a series of lectures at the Royal Society, just as he had in earlier years. Unfortunately for him, the results were more or less the same as before: the audience was largely unmoved by his theories, and they were soon upstaged anyway by those of Nicolaus Steno.

This was the 1690s, and Steno was long dead, of course. His stand-in for the renewed fossil debates was a scientific personality who could not have been more different—he was as extravagant in his geological speculations as Steno was circumspect; as bombastic and egocentric as Steno was demure and self-denying. He championed Steno's science, but loudly denied he was doing any such thing. The truth is, the version he presented was in such an odd and bastardized form that Steno might actually have wished to disown it, anyway.

John Woodward was a successful London physician who had clawed his way up from humble, if not impoverished, beginnings to wealth and prestigious memberships in the College of Physicians and the Royal Society. An independent, self-made man, he never attended university, and in all matters he sought no one's counsel but his own. Among other physicians he was sometimes known as "Don Bilioso" because of his idiosyncratic practice of prescribing vomiting as a cure for every ailment.

Learning his medicine through apprenticeship rather than formal schooling, Woodward found his first fossil at the age of twenty-five. On an outing in the country with his employer he had espied a seashell embedded in solid rock. It astonished him. He had never heard of such a thing, yet he quickly discovered the plowed fields were full of them. At once obsessed with the question of their origin, he resolved to "pursue it through all the remoter parts of the Kingdom."

Woodward became such an avid explorer of England's quarries, hillsides, and stream gullies that within four years he was boasting that he had already scoured nearly the whole country for fossils. It may have been the truth. In an incredibly short time, Woodward's personal collection of fossils had become one of the largest in London, attracting the attention of older and more established naturalists like Ray and Lister. Lister even sponsored Woodward's membership in the Royal Society. Before long, Woodward was soliciting specimens from correspondents as far away as North America, and cadging seashells from the likes of John Locke and Isaac Newton.

The only thing more remarkable than his growing cabinet of fossils was his ever-expanding ego. Woodward's reputation as a fossilist became such that dignitaries visiting London began to call on him to view his collection. His arrogance was legendary. A German count recalled being put off for days and then kept waiting for hours before being received by Woodward, who appeared in his dressing gown. The doctor's fossils were indeed impressive, said the count, but "he requires everyone to hang onto his words like an oracle, assenting to and extolling everything." Woodward's behavior was truly flabbergasting. "The most ridiculous thing of all is that he never ceases looking at himself in the mirrors, of which there are several in each room."

Yet with all his activity collecting and displaying his fossils, Woodward did not forget the puzzle that first had set him hunting seashells. In fact, he had solved it within two years of finding his first specimen. Shells in the earth had baffled the greatest philosophers for centuries, but to Woodward their explanation was "so plain," he told a friend, "he wonders no body thought of it sooner."

Woodward's theory was revealed to the world in a book entitled *An Essay Toward a Natural History of the Earth,* published in 1695. Against his mentor Lister and the prevailing opinion of the day, he scorned the earth's plastic powers, declaring that fossil shells were most certainly the real thing. His explanation for how they came to be found on land was, on the face of it, perfectly conventional. They were all left by Noah's Flood.

But the Flood as conceived by Woodward was different from everyone else's notion of Noah's Flood. He wove into his own theory some of the science of the only man in England he admitted was his intellectual equal—Isaac Newton.

The Flood was no mere deluge of water, he explained, but a temporary shutdown of Newtonian gravity. Such an event, decreed by God, would obviously cause all of the solid matter in the world to "instantly shiver into millions of Atoms and relapse into its primitive Confusion." The entire planet dissolved, turning into a single gigantic slurry of water and earth. After the Flood had done its deadly work, God reinstated gravitational forces, causing the whole "promiscuous Mass of Sand, Earth, Shells, and the rest" to collapse inward and resolidify.

This accounted for the earth's strata, he explained, because as the jumbled particles settled out from the flood waters, they would arrange themselves "according to the Order of their Gravity, those which are heaviest lying deepest in the Earth, and the lyter sorts . . . shallower." There could be no other explanation for the multitude of seashells buried so deeply in the earth.

But Woodward's "hasty pudding" theory of the earth, as the London wags called it, had some serious problems with its underlying science.

"When we consider how far it may agree with reason and common sense," the Welsh naturalist Edward Lhwyd confided in a letter to John Ray, "we find so many absurdities in it, that to me it seems scarce worth our consideration."

Actually, many of Woodward's critics (Lhwyd, for example) objected first and foremost to Woodward's identifying fossil shells as real shells—in situ growth was still the preferred theory. But here, Woodward had an advantage over his critics that Steno and Hooke lacked: His fossil collection was now acknowledged the richest in England. After his marathon fossil-hunting tramps across the countryside, no one could equal his firsthand knowledge of fossils—not even the redoubtable Martin Lister, his former mentor, who now found Woodward "impudent" and "troublesome to all mankind."

As for the perennially disturbing problem of extinct fossil species, Woodward solved it by flatly denying it. The species were all out there in the sea somewhere. Not even the peculiar ammonites were extinct. The Doctor "hath real Sea-shells of that kind now by him," wrote his friend John Harris in a pamphlet, "which I've more than once seen and compared with the fossil ones." Given Woodward's voracious collecting habits, and the confusion that still surrounded these fossils (what Harris probably saw were shells of the chambered nautilus), the claim seemed plausible enough to many people.

More serious were the criticisms of Woodward's account of strata. Like Woodward, both Steno and Hooke assumed that the flood was a real historical event, and that it accounted for some sedimentary layers. But neither believed it was the sole explanation for fossils and strata. And certainly they never called on such bizarre physics to produce fossil deposits.

Why, critics asked, if all the earth's solids were supposedly dissolved during the gravity shutdown, did seashells or, for that matter, Noah's ark survive the Flood intact? How could any sea life have lived through the flood in such a thick mass of earth and water? Did Noah have a fishpond aboard the ark?

John Arbuthnot, a mathematician and prominent London wit, found that, among other absurdities, Woodward's flood would need a source of water 450 miles deep just to dissolve the world into equal parts of earth and water; a conservative figure. He observed dryly: "The Doctor should have calculated the Proportions of his Drugs before he mix'd them."

Some of the criticisms got answers from Woodward. Organic materials like shell and wood did not dissolve in the zero-gravity floodwaters, he explained, because they consisted of interwoven fibers, which held them together while nonfibrous metal and stone was atomized. But most questions he deferred until he completed his magnum opus, which would fill in the details of the theory, barely outlined in his *Essay,* and vanquish all doubters.

There was one difficulty, however, that seemed impossible to explain away: Readers noticed that in the *Essay* many of the key scientific ideas for which Woodward claimed sole credit were not exactly new. They had been published twenty-five years earlier by none other than the "famous Mathematician and Philosopher" Nicolaus Steno. Yet nowhere in the book did Woodward acknowledge this.

He might have been able to claim that "great minds think alike" had it not been for John Arbuthnot's devastating pamphlet, *An Examination of Dr. Woodward's Account of the Deluge, &c., with a comparison between Steno's Philosophy and the Doctor's, in the Case of Marine Bodies dug out of the Earth.* Arbuthnot printed several pages of excerpts from both men's books, alternating between the authors so that the "reader may the better perceive in what they differ, and in what they agree."

It was plain that regarding fossils and strata, Woodward had cribbed almost every important point from Steno. He had copied some passages almost word-for-word. It was "in those parts that are most exceptionable," Arbuthnot observed sardonically, "the Doctor's Philosophy is different from Steno's." Steno had invoked no miracles, nor had he made such extravagant, unsupportable claims.

The best part of the book was when it argued for the organic origin of fossils. After so many years of relentless collecting, the doctor's expertise was undeniable. But out of all the arguments, Arbuthnot found only one to be original, and it was bogus at that. Woodward said fossils have the same medicinal properties as their biological counterparts. Arbuthnot not only doubted this was true, but doubted that Woodward had even tried any experiments to test it. Did the fussy and vainglorious fossil collector, who never allowed visitors to so much as touch the petrifactions in his cabinet, really feed any of his precious fossils to patients?

The idea that sediment grains settling through water sorted themselves by size to form layers was obviously lifted from Steno, too, but Woodward had badly garbled its meaning. Steno offered it to explain the arrangement of laminations within strata, not of the strata in the whole planet! It was preposterous to invoke miracles and overhaul the entire planet to explain such simple features. Steno had proved that rock strata were deposited by water, but he never said they were all dumped at once as part of a single miraculous event. "On the contrary," wrote Arbuthnot, "their Diversity and Order seems rather to persuade that they were compiled little by little and different times."

"God forbid I should limit Omnipotency," he concluded, "but as to Second Causes, I must remain an infidel till the Doctor's larger Work appears. I am sure Steno's Rule of forming the Strata is more compatible to the known Laws of Nature."

How did Woodward react to the damning evidence he had pilfered and distorted Steno's science?

He denied it absolutely. In typical fashion, he lashed out angrily at the "Impudence" of his accusers and wondered at the "Ignorance & Stupidity" of anyone who could believe such a charge. He insisted he owed nothing to Steno. "We hardly agree on any Thing," Woodward said, which was partly true, but evaded the point.

Even after the revelations by Arbuthnot, and despite his boorish personality, Woodward had many ardent supporters. John Harris, who had vouched for Woodward's living ammonites, was so taken with the theory and its vindication of the Biblical flood that he intimated that now only atheists and freethinkers could doubt the biological origin of fossil seashells. His fierce attacks on Woodward's critics earned him the title "the Anglican Inquisitor."

Others were calmer, if no less infatuated. Comparing Woodward to Newton, Richard Bentley, master of Trinity College at Cambridge, was moved to verse:

> Who nature's Treasures would explore,
> Her mysteries and Arcana know
> Must high as lofty Newton soar,
> Must stoop as delving Woodward low.

Partisans on both sides of the debate rarely expressed themselves in such poetic terms, however. "Methinks philosophers should not fall out about Shells and pebbles," said one Royal Society member early in the controversy. But in dealing with critics, Woodward seemed incapable of refraining from personal invective. At one gathering of the Society, the exchange over shells and pebbles became so heated that Martin Lister drew his sword. Only the interposition of another scientist between the two men, said an observer, prevented there being "Philosophical blood spilt."

As late as 1717—more than 20 years after the *Essay*'s publication—tempers still ran hot. A play lampooning Woodward in the character "Dr. Fossile" with his "Raree-Show of Oyster-shells and Pebble stones" ran successfully for seven nights, despite interruptions by hoots and catcalls from the packed house, but ultimately degenerated into a brawl between the pro- and anti-Woodward factions in the audience.

All along, Woodward promised that the magnum opus he had in the works would settle everything. But a former assistant told the story that this manuscript had sat atop a shelf in Woodward's study for years. It needed only "polishing," Woodward said, yet he refused to show it to anyone. One day, though, when Woodward stepped from the room, the assistant pulled down the book and opened it. The pages were mostly blank.

In Germany, Leibniz followed the Woodward controversy with a keen interest. His own geological ambitions were well-known within scholarly circles. In 1693, two years before Woodward's book came out, Leibniz published a brief abstract of his ideas in the journal *Acta Eruditorum,* which he had founded ten years earlier. "The terrestrial globe has undergone much greater changes than might be readily supposed," he announced, explaining that successive rock strata told the story.

After Woodward and his *Essay* became famous, the two men struck up a correspondence. Woodward, aware that Leibniz was the one person the rest of Europe considered Newton's intellectual peer, was uncharacteristically deferential. But to his mind, Leibniz's geology was insufficiently biblical, and he was unimpressed by Leibniz's knowledge of fossils. "I admire that he has no further Insight into those things, at this Time of day."

It was true, Leibniz had not devoted himself to fossil collecting as single-mindedly as had Woodward. And he was more concerned that his geology would confirm his own optimistic philosophy of the world than confirm the words of Scripture. Leibniz was convinced that the geological history of the earth, once revealed, would show the kind of progressive improvement his philosophy espoused. Fossils and strata were not relics of a single, earth-transforming calamity, as Woodward believed, but the preserved records of a harmoniously evolving world.

Leibniz hewed more closely to Steno in his thinking about rocks and fossils, and openly credited him for it. In the principles of stratification and the formation of rocks he saw "the seeds of a new science." "Natural Geography," as he thought it should be called, would have great practical benefits—mapping the earth's strata, for example, would help immeasurably the normally hit-or-miss art of mineral prospecting.

The draft of *Protogaea,* "The Primitive Earth," as Leibniz called his geological introduction to the Brunswick family history, was completed some time in the early 1690s. It did not rehearse the earth's past in any great detail, as he might have hoped. This turned out to be too daunting a project even for him. Continuing to dabble in geology, and pursuing his myriad other interests, Leibniz put off publishing *Protogaea.*

But in 1698 his indulgent employer, Duke Ernst August, died. The new duke, George Ludvig, was less patient and less interested in Leibniz's harebrained projects. All he wanted to know was: Where was the family history Leibniz was supposed to be writing? Leibniz had been working on it for more than ten years, but still had not made it past the year 1000. It was time to buckle down.

The duke sent him back to the genealogy books with a warning, but before long Leibniz was again writing philosophical dissertations and pursuing dozens of other ventures, even taking time out to investigate and write a report on a local farmer's talking dog.

But Leibniz's lucky star was fading. His relations with the duke deteriorated, and he was attacked by Isaac Newton and cronies for having plagiarized Newton's method of calculus. It is now known that Newton and Leibniz had developed their methods independently, but the feud between the two became vicious and personal, and Leibniz got by far the worst of it.

In the last years of his life, Leibniz grew increasingly isolated. Among the youthful sophisticates at court in Hannover he seemed ridiculous. The oversized wig and ornate ruffed clothing he wore were no longer the height of fashion—had not been for thirty years—and his expansive manner just made him seem all the more clownish. When George Ludvig left Hannover to be crowned King George I of England, he took his whole court with him—except Leibniz.

When he died in 1716, Leibniz still had not completed the Brunswick family history, nor had he cleared his name in the calculus imbroglio. In England, Newton supposedly gloated over breaking his rival's heart. In Hannover, Leibniz, universal genius and optimist of optimists, had no friends left to attend his funeral. He was mourned only by his secretary.

Protogaea was still unpublished. But was not destined to the same fate as Steno's lost dissertation. Leibniz had been a librarian after all, even if a nonpracticing one; he took good care of his papers. The manuscript was known, and when his collected philosophical works were published in 1749, *Protogaea* was among them.

Woodward died a decade after Leibniz and, according to his wishes, was buried in a place of honor near Isaac Newton in Westminster Abbey. In his will he bequeathed money to endow a professorship of paleontology at Cambridge University so his work could be carried on for posterity. It was the first academic position in history devoted to a science that was still barely in existence. The Woodwardian Chair has been held by many illustrious scientists ever since. Visitors to Cambridge can still see Woodward's collection, which remains intact in the original walnut cabinets with his own handwritten labels.

"I was sorry to hear of Dr. Woodward's death," wrote one contemporary. "He was a droll sort of philosopher, but one who had been at much pains and expense to promote natural knowledge. Some of his fossils were indeed very curious, though indeed he was the greatest curiosity of the whole collection."

Flamboyant and controversial, Woodward publicized the fossil question in a way Steno never could, and certainly never would, have dreamed doing. The *Essay* was an international bestseller, going through numerous editions in English, not to mention Latin, French, Italian, and German. If Woodward the theorist was not to be taken seriously, Woodward the phenomenon was not to be ignored. Blustering his way through the web of issues that had stymied more thoughtful naturalists such as John Ray, Woodward excited a wide new interest in fossils and strata. Ever increasing numbers of scientists were now inspired to study them closely, even if only to refute his obnoxious claims, perhaps hoping (vainly) to shut him up.

Overall, it's hard to say who played the biggest role in keeping Steno's ideas alive for the next generation of scientists—Leibniz, Woodward, or even the small enclave of naturalists in northern Italy, where memories of Steno as a scientist and religious figure were still fresh. Sadly, Robert Hooke, whose geological outlook was so similar to Steno's, never published his work. Not until after his death in

1703 were his lectures on fossils and earthquakes published, but even then, being in English, a language few continental scholars could read, they were quickly forgotten.

But now, as the new century was rising, more and more people were willing to believe that the seashells they found in the earth were indeed seashells. Before long, the seeds of Steno's new science, however scattered, would finally, and definitively, take root.

17

A NEW HISTORY
OF THE EARTH

Some drill and bore
The solid earth and from the strata there
Extract a register, by which we learn
That He who made, and reveal'd its date
To Moses, was mistaken in its age.

—WILLIAM COWPER, *THE TASK* (1785)

The eighteenth century was the century of Steno's triumph. His geological principles were finally put to work and began to reveal a new, unknown history of the earth. Oddly, this is also where he has traditionally disappeared from the history books.

The old story was that by the close of the seventeenth century, Steno and his geology had been forgotten. Not until more than a century later, after the British scientist James Hutton supposedly invented modern geology and opened up deep, geological time for exploration, were they "rediscovered." Only then, the story went, was Steno belatedly recognized as geology's founder. And even then, his forsaking science for religion, especially so soon after the birth of his new science, made him seem to some scientists and historians more like a deadbeat dad than founding father.

But the facts don't support the old story. Over the last few decades scholars have come to realize that during the hundred years between Steno and Hutton, geologists (as some were already starting to call themselves) were busy applying Steno's principles, mainly to the practical problems of mining and mineral surveys.

And as these scientists worked, it turns out, Steno's name was often on their lips. In many books on the new natural history he gets a respectful nod, in others his influence is obvious. During the 1700s, *De solido* was reprinted at least twice, once in Latin and once in French, the new universal language of scholars. Leibniz's *Protogaea*, finally published in midcentury, also in Latin and French, brought Steno's ideas to still more readers.

Their spread was uneven, and progress was halting, to say the least. But within a century the idea that the earth had its own history, separate from that of humans and not bound to the biblical tale, was surprisingly well-accepted not only by scientists, but the clergy as well. Yet this new past offered almost insurmountable challenges to the human imagination. Controversies could be as fierce as ever, with battle lines often strangely drawn.

By the first couple decades of the new century, Steno's arguments for the biological origin of fossil seashells finally prevailed, and the earth's rock strata were generally assumed to be deposits from water. Even rocks such as granite and basalt, now known to be igneous (hardened from molten rock) were considered to be sedimentary deposits.

But in the minds of many people, the structure of the earth's crust was tied to a single tumultuous event—the biblical Flood. Steno's principle of superposition was therefore a trivial idea. There was no history to be read into it besides the clearing of turbid flood waters after the main event. Even the tilting and folding of originally horizontal rock strata was assumed to be the consequence of the crust's collapse immediately following the Flood. Woodward and his disciples were adamant that since the Flood, no other changes had taken place on the face of the earth. Even many of Woodward's critics agreed on this point.

But even as Woodward preened and took bows for the international success of his radical Flood theory, it was already being undermined by studies in the field. He and his followers could not explain away the fact that the fossils in rock strata were *not* arranged in order of their specific gravity. And anyone who took the time to stir up a bucket of muddy water could see that only when the water became still would the finest material settle out. How, then, could the layers of silt and clay interleaved with the sand and gravel have formed during the supposed violence of the Flood?

But these were old arguments, already talked to death by gentleman scientists frustrated by the slippery philosophical and theological issues that were involved. Ultimately, the evidence turned the tide against the biblical Flood as a catch-all explanation for fossil shells out of more pragmatic concerns.

In 1720, a report with the bland title *Remarks on some fossil shells of Touraine and their uses* was submitted to the Paris Academy of Sciences. True to his title, the author, a chemist by the name of René Réaumur, devoted most of his report to a factual description of the *faluns*, thick sedimentary layers made almost entirely of broken shells, that underlay much of the region of Tours, explaining how the local farmers quarried the shells to use as a soil conditioner and giving the locations of the richest deposits. It contained also a simple but fatally damaging argument against not only the flood theory, but the biblical timescale as well.

In noting the wide extent of the *falun* layers and their thicknesses, up to seven meters, Réaumur pointed out that a flood that supposedly had lasted less than a year would not have left such a thick, evenly layered deposit. It seemed more likely to Réaumur that Tours had once been occupied by an arm of the sea.

He didn't know how long ago the superimposed shell layers had been deposited, nor how long it had taken for them to be laid down, though he suspected it was a "long time." But from the slowness of historical changes in the French coastline, he estimated that it would take thirty to forty centuries for the sea to recede from Tours to its present position.

Compared to the geographic changes implied by seashells farther inland or up in high mountains, the retreat of the sea from Tours was a minor shift. And it was only a single event, in some areas the strata implied several such invasions. Yet Réaumur was suggesting that it took half the time the Bible allowed for all the world's history!

Ever since Steno had published *De solido,* a number of scientists had quietly wondered about its implications for biblical chronology, which was quickly unraveling. By 1701, when James Ussher's 4004 B.C. date of Creation was first printed in the margins of the King James Bible, to become Anglican orthodoxy, it was already in trouble.

The old chronology was being attacked on several fronts. One of the first shocks had been the discovery of the New World with its native population. Where did *they* come from? To explain this, and to reconcile inconsistencies in the Genesis text, Isaac de Lapeyrère proposed that there had been people on earth before Adam. The Pre-Adamites, as he called them, were created during the first phase of Creation, which is recounted in Genesis 1. Adam and Eve did not appear until the next phase, told in Genesis 2. An untold number of years, he claimed, must have passed between the two episodes of Creation.

Lapeyrère's theory was quickly suppressed by both Protestant and Catholic authorities, but doubts lingered as more disquieting news came from overseas. The civilizations in the Americas conquered by the Spanish claimed to have histories that went back tens of thousands of years into the past. Missionaries to China found out that Chinese had chronicled more than four millenia of uninterrupted history. Reckoning by Ussher's timetable, that took them back beyond the supposedly universal Flood.

As if they didn't have enough problems from these external threats, biblical chronologists shook up the orthodoxy with their own internal controversies.

Other chronologies besides Ussher's famous version, based variously on the different Hebrew, Greek, and Latin texts, vied for respectability. Isaac Newton was a dedicated biblical chronologist, spending more than a decade on his own idiosyncratic timetable. One seventeenth-century priest compiled a catalog of more than seventy separate biblical chronologies known to him. By the early eighteenth century, the number had swelled to the hundreds, no two exactly the same. The discrepancies could be glaring. Was the Flood in the year 1656 after Creation? Or the year 2256? Or the year 3882?

One could hope that the biblical scholars would eventually resolve their differences, and many dismissed the heathen chronologies as the unreliable fantasies of barbarians. But the lesson of the Galileo affair had not been lost on the Catholic Church or its Protestant counterparts: to deny a theory in the face of hard empirical evidence was a losing game.

And petrified seashells were nothing if not hard. To some, in fact, they epitomized the kind of hard evidence one could glean from nature. The "oldest library in the world" had too long been left unread, and fossil seashells were icons of a new literacy and an expansive new vision of the world.

"Who will be so obstinate as to refuse the truth arising so clearly from this discovery?" wrote Benoît de Maillet, a French diplomat who traveled extensively through Egypt and the Middle East, observing its rock strata and abundant fossil seashells. While some were inclined to stretch the biblical timescale, he wanted to discard it entirely. The world must be hundreds or thousands of millions of years old, or perhaps even eternal.

"The opinions of six thousand years will pass like the aberration of one day," wrote another Frenchman, Nicolas Boulanger. "The immense heaps of shells on the tops of mountains today as far distant from the seas as possible" spoke of a world "Thousands of centuries" old. "These facts, unknown to the vulgar, but well known to all who observe nature, force the physical scientist to recognize that all the surface of our globe has changed; that it has had other seas, other continents, another geography."

Boulanger, a civil engineer who made his observations of fossils while working in various parts of France, wrote little, but he brought the message of seashells to Paris, where he mingled with the iconoclastic philosophes Denis Diderot, Baron D'Holbach, and Jean Jacques Rousseau. A geological evangelist, he burned himself out and died at the age of thirty-seven. By then, said Rousseau, his "exalted imagination no longer saw anything in nature but shells, and he ended up really believing that the universe was nothing but shells."

Neither Boulanger nor De Maillet published their heterodox ideas—it was still better to be discreet, even though their ideas were shared by many. But their work circulated widely in manuscript form, and after thirty years De Maillet's was published posthumously, with his authorship thinly disguised by reversing the letters of his name. A kind of geological "romance," *Telliamed, Conversations between an Indian Philosopher and a French Missionary on the Diminution of the Sea*, theorized that an original universal ocean had progressively dwindled over the eons, leaving behind mountains of sediments and fossil shells.

One of the first to go public with daring new geological theories was George Louis Leclerc, Compte de Buffon. The same year *Telliamed* was published, he proposed in his bestselling *Natural History* that the seas were not shrinking, only migrating. Left behind by the rotating earth, the water crept westward little by little each year. Submarine mountains heaped up by ocean currents became mountains on land, and the land, in turn, became the ocean floor. He ignored the Flood completely.

There were others, however, to whom rejecting the Flood meant rejecting the idea of organic fossils altogether and going back to wooly explanations of the previous century. This reaction was not limited to religious conservatives worried about the authority of Scripture. To some scientific-minded types, theories of universal oceans and migrating seas seemed just as fanciful as Bible tales about Noah and the ark.

"These imaginings dishonor physics; such charlatanry is unworthy of history," said their most pungent critic, Voltaire. Discrediting hypotheses and philosophical systems based on "seashells" became a personal crusade.

The fact was, theorists often did mix some fantasy in with their science—that was part of their appeal. Along with his theory of the shrinking sea, de Maillet proposed a kind of theory of evolution, in which all terrestrial creatures were the transformed descendants of marine life left stranded on dry land. Birds were formerly flying fish, humans mermaids and mermen. Boulanger, like many of the free-thinkers who jumped on the old-earth bandwagon, made the anthropocentric assumption that human beings were as eternal as the earth. He fancifully described their terror as they witnessed the floods and earthquakes that created mountains and rock strata, after which they "wandered through the ruins of the world at the mercy of all the torments that seemed to persecute them."

"Philosophers want to see great changes on the stage of the world," said Voltaire, "just as the people want them in spectacles." A simpler explanation for the many shells in the mountains, he joked, was that they were the remains of snacks dropped by travelers crossing the mountains, or the shell badges lost by wandering pilgrims.

Like other deists, Voltaire was committed to the Newtonian vision of the universe as a perfectly designed machine, operating on logical, mathematical principles. The kind of upheavals proposed by the geologists, even ones more sober than de Maillet and Boulanger, had no place in this universe.

The geology Voltaire admired was Kircher's. He had read *Mundus subterraneus* as a student and particularly seized on Kircher's thesis that mountains were permanent and necessary features of the planet. This made genuine marine fossil in the mountains an impossibility.

The idea of spontaneous generation had never completely died among some of the more freethinking and iconoclastic members of the intelligentsia, who liked it for the same reason that conservative Christians had come to hate it: It suggested that Nature had its own potency, separate from God.

But to Voltaire, spontaneous generation appealed because it could put shells in the mountains without disturbing the rational order of the globe. There was no other conclusion a logical mind could draw.

"It always astonishes me," he said of ammonites, "that some refuse to allow that the earth produces these stones."

He specifically rejected Réaumur's conclusions about the shell layers. He said he knew of a château in Tours where the ground was so rich in fossil shells that one could almost watch them "vegetate" on the spot.

A few years after Voltaire's death, Thomas Jefferson, who had a long-standing interest in seashells on mountaintops because of fossils he'd seen in the Blue Ridge Mountains, traveled to Tours in search of the same château. A local man swore it was true, but Jefferson left unconvinced. The "origin of shells in high places," he concluded, might just be one of those questions "beyond the investigation of human sagacity."

Jefferson shared Voltaire's deistic beliefs, and though he refused to go out on a limb concerning fossils he never could accept that the earth might have a significant geologic past. Months before his death he wrote: "The dreams about the modes of creation, inquiries whether our globe has been formed by the agency of fire or water, how many millions of years it has cost Vulcan or Neptune to produce what the fiat of the Creator would effect by a single act of will, is too idle to be worth a single hour of any man's life."

But men already spent years doing just that. Beginning in the 1750s, there was a rush to survey the earth's strata and map its mineral riches. The driving forces behind their geological investigations were not just religious or philosophical: They were political and economic. The developing industrial revolution was raising the economic stakes, and every country wanted to know what mineral resources it had within its borders.

In 1760, an Italian mining expert named Giovanni Arduino circulated a letter among his colleagues. In it he laid out a scheme for classifying the strata of the Italian Alps.

He separated them into three main groups, which he called Primary, Secondary, and Tertiary. The Primary strata, at the bottom of the pile, were tilted and lacked fossils. Next, were the Secondary strata, which were tilted but *had* fossils in them. On top were the Tertiary strata, which were horizontal and also contained fossils. He added a fourth class, Quaternary, for the sands and gravels that covered the bedrock in the nearby Po River Valley.

Arduino had been inspired by "the celebrated supposition of Leibniz and other learned men to the effect that our earth does not possess its original aspect, and that that which it now has is a result of innumerable complicated effects of fire and water." His system, though, was a direct descendent of Steno's. Reading *De solido* and visiting Steno's old haunts in Tuscany, he took Steno's two cycles of sedimentary strata and divided them into four.

Unlike Steno, however, he did not try to find any connection between his rock strata and the Bible. Nor did he assume that the primary rocks at the bottom of his sequence dated from the beginning of the world. Arduino claimed only that his strata marked "distinct epochs in the history of the earth."

Scientists following him would rename his Primary and Secondary divisions the Paleozoic and Mesozoic systems. Arduino's Tertiary and Quaternary divisions still stand. Together, his four divisions encompass nearly six hundred million years of the earth's history, from the first evolutionary explosion of marine life to sediments laid down in recent human history.

Arduino had in fact noted the idiosyncrasies of fossil species that a century earlier had confused Martin Lister and intimidated John Ray. Fossils in his older Secondary strata were "unrefined and imperfect," but those in the Tertiary above were "very perfect and wholly similar to those that are seen in the modern sea."

"As many ages have elapsed in the elevation of the Alps," he concluded, "as there are races of organic fossil bodies embedded within the strata."

Neither Arduino nor any of his contemporaries had a way to measure the length of these "ages." Not until the beginning of the twentieth century did a method become available. Yet over the next century geologists got surprisingly far with little more to guide them than what Arduino had: Steno's stratigraphic principles.

No one could put a precise number on the vast age of the world, but by 1800 all but a few scientists believed it could be counted in millions, even billions of years. Those few geologists still looking for the remains of Noah's Flood did not climb mountains in search of petrified seashells. Instead, they looked in the layers of sand and gravel covering the surface in certain parts of Europe, deposits that later turned out to be from Ice Age glaciers, not a Scriptural Flood.

Following the geologists' exploits, egging them on, and even joining in were the poets and writers of the new Romantic movement. Critics had long sneered at geological theories of unfathomable past as mere "romances," but this hardly detracted from their appeal to the Romantics, who saw geologists as kindred spirits.

In fact, the geological academy in the German mining town of Freiburg was as much a seedbed of Romanticism as science. The poet Novalis, among others, took classes and worked in the mines. Goethe studied there and worked for a while as a mining administrator. Traveling to Italy for a tour of writing and nature study, Goethe brought with him his copy of Steno.

Except among the most rigidly literal-minded, the biblical timescales expounded by Ussher and others were abandoned, or, more often, stretched. In the early nineteenth century, Pope Pius VII told Catholics it was permissible to view the days of Creation as indeterminate "periods" rather than literal twenty-four-hour days.

When Darwin proposed his theory of evolution in mid-century, it was the theory itself, with the implied bestial origin of humans and lack of divine control, that drew fire from the clergy, not the huge expanse of time it demanded. Until the discovery of radioactivity gave a method to directly measure the ages of rocks, the most vehement critics of geological time were physicists, not preachers.

What would Steno have thought of the change in timescale his *De solido* wrought? Several of his contemporaries—Ray, Hooke, and Leibniz in particular—pondered, with varying degrees of alarm, the implications of fossils and strata for the biblical timescale. But so far as is known, Steno was silent.

In *De solido*, he compares his history of Tuscany with the biblical account and finds agreement, but mostly declines to speculate on the "succession of years of unknown duration" over which fossils accumulated.

Steno may not have had the same misgivings as Ray and other English Protestants. As a Catholic, he may have been less concerned about conforming to the Ussherian chronology. Ussher, after all, was an Anglican Archbishop and an avowed enemy of Rome. Possibly

Steno was satisfied to give the world a linear and roughly biblical history as proof against the eternalist heresy.

Unlike the others, he never questioned the adequacy of the Scriptural timescale. In fact, he raised the opposite concern—the possibility of *too much* time. Pointing out that shells were found in the walls of the Etruscan city of Volterra, known from historical records to have been built three thousand years earlier, and that perfectly preserved shells were also in the rock strata beneath the city, he reassured his readers that 4,000 years was not *too long* for sea shells to survive buried in the earth. He was pushing to expand their mental time limits to multiple thousands of years.

How far he would have pushed his own limits cannot be known. It is tantalizing to speculate that he might have broached the issue in his lost dissertation, but the manuscript has never been found. The eighteenth-century naturalist Targioni Tozzetti, who rescued the papers of the Cimento from oblivion, searched the Medici archives. He found Steno's fossil and mineral collection with a hand-written index of its specimens, but no other geological papers. If they were indeed sent to Copenhagen, most likely they were destroyed in a fire that ravaged the university in 1728. Yet they may still be in Florence, stashed away somewhere in an old archive, library, or museum cabinet.

Epilogue

THE SAINT

Men's curiosity searches past and future
And clings to that dimension. But to apprehend
The point of intersection of the timeless
With time, is an occupation for a saint—
No occupation either, but something given
And taken, in a lifetime's death in love,
ardour and selflessness and self-surrender.
　　　　　　　　　—T.S. ELIOT, *THE DRY SALVAGES* (1941)

If by some fiat I had to restrict all this writing to one sentence, this is the one
I would choose: The summit of Mount Everest is marine limestone.
　　　　　—JOHN MCPHEE, *ANNALS OF THE FORMER WORLD*
　　　　　　　　　　　　　　　　　　　　　　　　　(1998)

Nicolaus Steno solved a puzzle that had nagged the human mind for centuries. How do seashells get on mountaintops? It took many years for his work to bear fruit, but its ultimate impact is hard to overestimate.

He showed that the earth had a history, revealed in its own rocks. As a result, the static world assumed by both the scientists and churchmen of his day gave way to an evolutionary one. And with that idea came unlimited possibilities.

The earth is now known to be 4.6 billion years old. Seashells on mountaintops are no longer a mystery. The science of geology employs thousands of people in academia, government, and industry. In searching for oil and mapping the structure of bedrock, geologists use Steno's principles every day. Students of geology quickly learn to take them for granted. Out of these seventeenth-century ideas grew the modern concept of deep time, not to mention plate tectonics, evolution, global climate change, and dinosaurs. It all would greatly surprise Steno.

What might seem surprising to us looking back at Steno today is his status as a figure of both scientific innovation and staunch religious piety. Despite *De solido*'s clear connection to the volatile issue of interpreting Scripture, Steno's science was never criticized by his contemporaries on religious grounds. There is no evidence that the Church objected to anything Steno wrote, or pressured him in any way. Both the hyper-orthodox Cosimo de Medici and the progressive Leibniz encouraged Steno to continue his geological research. As for Steno, even at the height of his religious zeal, he never disowned his science, never showed any concern that it might infringe on his faith.

Conditioned by the familiar story of Galileo's persecution by the Catholic Church and by the modern-day clash between scientists and Protestant fundamentalists over evolution in the classroom, we often assume antagonism between religion and science is inevitable. But as much as their methods and ideals differ today, over the history of both there has been easily as much cross-fertilization as conflict. Until very recently, religious and scientific arguments were advanced by both sides in every important scientific controversy. Too often what filters down to us in the history books are the scientific arguments of the winners and the religious arguments of the losers. Thus the picture of a long-standing rift between the two.

Steno credited his scientific career for freeing him to explore different religious beliefs. Scientists were more likely than other intellectuals of his time to mingle amicably with colleagues of contrasting faiths. It was this, he said, that opened the door to religious conversion.

And for Steno science, like religion, was an occasion not for arrogance, but humility. He could not guess the expanse of time his science would reveal, but he never flinched at the vastness and imponderablity of the universe, or the comparative frailty of the human species.

Steno's fame as a churchman never matched his fame as a scientist; he gave that up when he gave up the material trappings of the Medici court for a life of self denial. But Steno was not forgotten. Ultimately he achieved the greatest recognition he could have wanted.

In 1722, Jacob Winslow, Steno's great-nephew and an accomplished anatomist in his own right, wrote a short biography of his famous kinsman. It was published in a book called *Lives of the Saints for each Day of the Year*, which was published in Paris and had a section for prospective saints. Like Steno, Winslow had converted from Lutheranism to Catholicism. In fact, he attributed his conversion to Steno's intercession.

Two hundred years later, in 1938, the tricentenary of Steno's birth, a group of Danish pilgrims appealed to Pope Pius XI to have him canonized as a saint. The pope directed them to begin gathering the necessary proof of Steno's sanctity and "heroic virtue."

The next step came in 1953. Steno's coffin was found in the crypt of San Lorenzo and the lid removed to identify the remains. The skeleton, missing its cranium, was dressed in bishop's robes, with a crosier laid alongside.

Then, in a solemn procession, his body was carried through the streets of Florence and relocated to a chapel off the nave of San Lorenzo, renamed "Capella Stenoniana." Steno was placed in a fourth-century Christian sarcophagus, excavated some years earlier from Arno River sediments, and donated by the Italian government. He is still there today.

After Steno's reinterment, a committee appointed by the church deliberated for twenty years, hearing testimony of possible miracles and instances of healing through Steno's intercession. They published their official report, the *positio*, in 1974, but a final decision from the pope to go ahead with the canonization had to wait another ten years. The committee had confirmed only one miracle: After praying to Steno, a cancer patient had an apparently spontaneous recovery. This was enough for beatification, but not full canonization.

In 1988, before some twenty thousand worshipers in Saint Peter's Basilica, Pope John Paul II said a mass of beatification. He praised Steno's science and his saintly life. Whether by design or coincidence, Steno was beatified on October 23, the same date Bishop James Ussher had chosen for the creation of the world.

Geologic Time

(information from the Geological Society of America; not to scale)

Phanerozoic Eon (543 million years ago to present)	Cenozoic Era (65 million years ago to present)	Quaternary Period (1.8 million years ago to present)	Recent Epoch (10,000 years ago to present)
			Pleistocene Epoch (10,000 years to ca. 2 million years ago)
		Tertiary Period (65 to 1.8 million years ago)	Pliocene Epoch (5 to ca. 2 million years ago)
			Miocene Epoch (24 to 5 million years ago)
			Oligocene Epoch (34 to 24 million years ago
			Eocene Epoch (55 to 34 million years ago)
			Paleocene Epoch (65 to 55 million years ago)
	Mesozoic Era (248 to 65 million years ago)	Cretaceous Period (144 to 65 million years ago)	
		Jurassic Period (206 to 144 million years ago)	
		Triassic Period (248 to 206 million years ago)	
	Paleozoic Era (543 to 248 million years ago)	Permian Period (290 to 248 million years ago)	
		Carboniferous Period (354 to 290 million years ago)	
		Devonian Period (417 to 354 million years ago)	
		Silurian Period (443 to 417 million years ago)	
		Ordovician Period (490 to 443 million years ago)	
		Cambrian Period (543 to 490 million years ago)	
Proterozoic Eon (2.5 billion years ago to 543 million years ago)			
Archean Eon (3.9 to 2.5 billion years ago)			
Hadean Eon (4.6 to 3.9 billion years ago)			

Sources

A full-blown scholarly biography of Steno in English has yet to be written. This book is no more than a start. Because I was mainly interested in his contribution to science as a geologist, I had to leave out many details of his careers as an anatomist and a priest.

Most English-language accounts of Steno's life are literally hagiographies, written in connection with his beatification. Not surprisingly, they smooth over the complexities of his personality. Of Steno's own writings that have been translated, aside from his scientific work, most are dominated by religious pieties and aphorisms that give little hint of the genuineness and warmth that his contemporaries found so appealing, and that made him an effective proselytizer for his faith. Danish speakers can get a more rounded picture of their illustrious countryman: In addition to several in-depth biographies written in Danish, Steno's life has been the subject of at least one novel (enlivened by a fictitious secret romance with a gypsy) and a verse play.

Steno's collected scientific works were published in 1910 (*Nicolai Stenonis Opera Philosophica,* edited by Vilhelm Maar, Copenhagen, 1910), his collected sermons and theological works in the nineteen forties (*Nicolai Stenonis Opera Theologica,* edited by Knud Larsen and Gustav Scherz. Copenhagen and Friburg, 1941–1947) and his letters in 1952 (*Nicolai Stenonis Epistolae et epistolae ad eum datae,* edited by Gustav Scherz and Joanne Ræder. Copenhagen and Friburg, 1952). His writings are mostly in Latin and most are still untranslated into

English. Most of the quotations from Steno's geological work I use in this book are those of Alex J. Pollack in *Steno: Geological Papers* (cited below).

This book is aimed at general readers, and is not intended to be a scholarly work. For this reason I have not included endnotes or compiled an exhaustive reference list. The books and articles listed below were my primary sources of information, and should be more than enough for readers wishing to learn more about Steno and the history of geology.

STENO

Cioni, Raffaello. *Niels Stensen: Scientist—Bishop.* Translated by Genevieve M. Camera. New York: P.J. Kenedy & Sons, 1962.

Garboe, Axel. "Niels Stensen's (Steno's) Lost Geological Manuscript." *Meddelelser fra Dansk Geologisk Forening* 14 (1960): 243–246.

Gould, Stephen Jay. "The Titular Bishop of Titiopolis." *Natural History* 90 (1981): 20–24.

Herries Davies, Gordon L. "The Stenonian Revolution." In *Rocks, Fossils and History,* edited by Gaetano Giglia, Carlo Maccagni, and Nicoletta Morello. Florence: Edizioni Festina Lente, 1995.

Hsu, Kuang-tai. "Nicolaus Steno and His Sources: The Legacy of the Medical and Chemical Traditions in His Early Geological Writings." Ph.D. Dissertation, University of Oklahoma, 1992.

Kardel, Troels. "Steno: Life, Science, Philosophy." *Acta Historica Scientiarum Naturalium et Medicinalium* 42 (1994): 1–159.

Kardel, Troels, ed. "Steno on Muscles." *Transactions of the American Philosophical Society* 84 (1994): 1–57.

Moe, Harald. *Nicolaus Steno, an Illustrated Biography: His Tireless Pursuit of Knowledge, His Genius, His Quest for the Absolute.* Translated by David Stoner. Copenhagen: Rhodos, 1994.

Pontifica Academia Scientarium. *Blessed Niels Stensen and His Memorial Plaque in the Pontifical Academy of Sciences.* Rome: Academia Civitate Vaticana, 1989.

Poulsen J.E., and E. Snorrason, eds. *Nicolaus Steno 1638–1686. A Reconsideration by Danish Scientists.* Gentofte, Denmark: Nordisk Insulinlaboratorium, 1986.

Scherz, Gustav, ed. *Nicolaus Steno and His Indice.* Copenhagen: Munksgaard, 1958.

———. "Niels Stensen's First Dissertation." *Journal of the History of Medicine and Allied Sciences* 15 (1960): 247–264.

———, ed. *Steno and Brain Research in the Seventeenth Century. Proceedings of the International Historical Symposium on Nicolaus Steno,* Held in Copenhagen, 18–20 August 1965. *Analecta Medico-Historica.* 3. pp. 1–302. Oxford: Pergamon Press, 1968.

———, ed. *Steno: Geological Papers.* Translated by Alex J. Pollock with Latin Text and Notes. Odense, Denmark: Odense University Press, 1969.

———, ed. *Dissertations on Steno as Geologist.* Odense, Denmark: Odense University Press, 1971.

Scherz, Gustav and Peter Beck. *Niels Steensen (Nicolaus Steno), 1638–1686: The Goldsmith's Son from Copenhagen Who Won World Fame as a Pioneering Natural Scientist but Who Sacrificed Science to Become a Celebrated Servant of God.* Copenhagen: Royal Danish Ministry of Foreign Affairs, 1988.

Steno, Nicolaus. *The Prodromus of Nicolaus Steno's Dissertation Concerning a Solid Body Enclosed by Process of Nature Within a Solid.* An English Version with an Introduction and Explanatory Notes by

John Garrett Winter, University of Michigan. New York: The
Macmillan Company, 1916.

——————. *Lecture on the Anatomy of the Brain.* Edited by Gustav Scherz
with anatomical annotations by Adolph Faller. Copenhagen: Nyt
Nordisk Forlag, Arnold Busck, 1965.

Ziggelaar, August, ed. *Chaos: Niels Stensen's Chaos-manuscript Copenhagen, 1659.* Complete Edition. Copenhagen: Danish National
Library of Science and Medicine, 1997.

GENERAL

Adams, Frank Dawson. *The Birth and Development of the Geological Sciences.* 1938. Reprint edition. New York: Dover, 1954.

Albritton, Claude C., Jr. *The Abyss of Time: Changing Conceptions of the
Earth's Antiquity after the Sixteenth Century.* San Francisco: Freeman
Cooper Publishers, 1980.

Ariew, Roger. "A New Science of Geology in the Seventeenth Century?" *Revolution and Continuity: Essays in the History and Philosophy
of Early Modern Science,* edited by Peter Barker and Roger Ariew.
Washington, D.C.: Catholic University of America Press, 1991.

Barr, James. "Why the World was Created in 4004 B.C.: Archbishop
Ussher and Biblical Chronology." *Bulletin of the John Rylands University Library* 67 (1985): 575–608.

Bedini, Silvio A. *Thomas Jefferson: Statesman of Science.* New York:
Macmillan, 1990.

Beretta, Marco. "At the Source of Western Science: the Organization of
Experimentalism at the Accademia Del Cimento (1657–1667)."
Notes and Records of the Royal Society of London 54 (2000): 131–151.

Carozzi, Marguerite. "Voltaire's Attitude Toward Geology." *Archives des Sciences* 36 (1983): 1–145.

Cochrane, Eric W. *Florence in the Forgotten Centuries, 1527–1800: A History of Florence and the Florentines in the Age of the Grand Dukes.* Chicago: University of Chicago Press, 1973.

Cohn, Norman. *Noah's Flood: The Genesis Story in Western Thought.* New Haven: Yale University Press, 1996.

Cole, F.J. *History of Comparative Anatomy.* London: Macmillan, 1944.

Collier, Katharine Brownell. *Cosmogonies of Our Fathers: Some Theories of the Seventeenth and the Eighteenth Centuries.* 1934. Reprint edition. New York: Octagon Books, 1968.

Cottingham, John, ed. *The Cambridge Companion to Descartes.* Cambridge: Cambridge University Press, 1992.

Dean, Dennis R. "The Age of the Earth Controversy: Beginnings to Hutton." *Annals of Science* 38 (1981): 435–456.

Desmond, Adrian J. "The Discovery of Marine Transgressions and the Explanation of Fossils in Antiquity." *American Journal of Science* 275 (1975): 692–707.

Drake, Ellen Tan. *Restless Genius: Robert Hooke and his Earthly Thoughts.* Oxford: Oxford University Press, 1996.

Ellenberger, François. *History of Geology, Volume 1: From Ancient Times to the First Half of the XVII Century.* Translated by Lt. Col. R. K. Kaula. Rotterdam and Brookfield, Vermont: A.A. Balkema, 1996.

———. *History of Geology Volume 2: The Great Awakening and its First Fruits, 1660–1810.* Edited by Marguerite Carozzi. Rotterdam and Brookfield: A.A. Balkema, 1999.

Findlen, Paula. "Jokes of Nature and Jokes of Knowledge: the Playfulness of Scientific Discourse in Early Modern Europe." *Renaissance Quarterly* 43 (1990): 293–331.

Findlen, Paula. *Possessing Nature: Museums, Collecting, and Scientific Culture in Early Modern Italy.* Berkeley: University of California Press, 1994.

Gillispie, Charles Coulston, ed. *Dictionary of Scientific Biography.* 18 vols. New York: Scribner, 1975–1985.

Gohau, Gabriel. *A History of Geology.* Translated by Albert V. Carozzi and Marguerite Carozzi. Revised edition. New Brunswick: Rutgers University Press, 1990.

Gottdenker, Paula. "Francesco Redi and the Fly Experiments." *Bulletin of the History of Medicine* 53 (1979): 575–592.

Gould, Stephen Jay. *Time's Arrow, Time's Cycle: Myth and Metaphor in the Discovery of Geological Time.* Cambridge: Harvard University Press, 1987.

Haber, Francis C. *The Age of the World: Moses to Darwin.* Baltimore: Johns Hopkins University Press, 1959.

Hamm, Ernst P. "Knowledge from Underground: Leibniz Mines the Enlightenment." *Earth Sciences History* 16 (1997): 77–99.

Herries Davies, Gordon L. *The Earth in Decay: A History of British Geomorphology, 1578–1878.* New York: American Elsevier Publishing Company, 1969.

Hibbert, Christopher. *The House of Medici: Its Rise and Fall.* New York: Morrow Quill, 1980.

Ito, Yushi. "Hooke's Cyclic Theory of the Earth in the Context of Seventeenth Century England." *British Journal for the History of Science.* 21 (1988): 295–314.

Jolley, Nicholas, ed. *The Cambridge Companion to Leibniz.* Cambridge: Cambridge University Press, 1995.

Kubrin, David Charles. "Newton and the Cyclical Cosmos." *Journal of the History of Ideas* 28 (1967): 325–346.

Laudan, Rachel. "From Mineralogy to Geology: The Foundations of the Earth Sciences, 1660–1830." *Science and Its Conceptual Foundations*, edited by David L. Hull. Chicago: University of Chicago Press, 1987.

Levine, Joseph M. *Dr. Woodward's Shield: History, Science, and Satire in Augustan England*. 1977. Reprint, with new preface. Ithaca: Cornell University Press, 1991.

Lewis, C.L.E. and S.J. Knell, eds. *The Age of the Earth from 4004 BC to AD 2002*. London: Geological Society of London Special Publication 190, 2001.

Magruder, Kerry. "Theories of the Earth from Descartes to Cuvier: Natural Order and Historical Contingency in a Contested Textual Tradition." Ph.D. Dissertation, University of Oklahoma, 2000.

Middleton, W. E. Knowles. *The Experimenters: A Study of the Accademia del Cimento*. Baltimore: Johns Hopkins University Press, 1971.

Montgomery, Scott L. "The Eye and the Rock: Art, Observation and the Naturalistic Drawing of Earth Strata." *Earth Sciences History* 15 (1996): 3–24.

Morello, Nicoletta. "*De Glossopetris Dissertatio:* The Demonstration by Fabio Colonna of the True Nature of Fossils." *Archives Internationales d'Histoire des Sciences,* 31 (1981): 63–71.

Oldroyd, David R. *Thinking About the Earth: A History of Ideas in Geology*. Cambridge: Harvard University Press, 1996.

Oldroyd, David R. and J. B. Howes. "The First Published Version of Leibniz's *Protogaea*." *Journal of the Society for the Bibliography of Natural History* 9 (1978): 56–60.

Porter, Roy. "John Woodward: 'A Droll Sort of Philosopher.'" *Geological Magazine* 116 (1979): 335–343.

Rappaport, Rhoda. "Geology and Orthodoxy: The Case of Noah's Flood in 18th-Century Thought." *British Journal for the History of Science* 11 (1978): 1–18.

Rappaport, Rhoda. *When Geologists Were Historians, 1665–1750.* Ithaca: Cornell University Press, 1997.

Rosenberg, Gary D. "An Artistic Perspective on the Continuity of Space and the Origins of Modern Geologic Thought." *Earth Sciences History* 20 (2001): 127–155.

Rossi, Paolo. *The Dark Abyss of Time: The History of the Earth and the History of Nations from Hooke to Vico.* Translated by Lydia G. Cochrane. Chicago: University of Chicago Press, 1984.

Rudwick, Martin J. S. *The Meaning of Fossils: Episodes in the History of Palaeontology.* 2d ed. Chicago: University of Chicago Press, 1976.

Schneer, Cecil J. "The Rise of Historical Geology in the Seventeenth Century." *Isis* 45 (1954): 256–268.

Schneer, Cecil J., ed. *Toward a History of Geology.* Proceedings of the New Hampshire Inter-Disciplinary Conference on the History of Geology, September 7–12, 1967. Cambridge: Massachusetts Institute of Technology Press, 1969.

Shapin, Steven. *The Scientific Revolution.* Chicago: University of Chicago Press, 1996.

Steneck, Nicholas H. *Science and Creation in the Middle Ages.* South Bend: Notre Dame University Press, 1976.

Toulmin, Stephen and June Goodfield. *The Discovery of Time.* 1965. Reprint edition. Chicago: University of Chicago Press, 1982.

Tuveson, Ernest Lee. "Swift and the World-Makers." *Journal of the History of Ideas.* 11 (1950): 54–74.

Vaccari, Enzio. "Mining and Knowledge of the Earth in Eighteenth-Century Italy." *Annals of Science* 57 (2000): 163–180.

Westfall, Richard S. *The Construction of Modern Science: Mechanisms and Mechanics.* New York: Wiley, 1971.

Willey, Basil. *The Seventeenth Century Background.* 1935. Reprint edition. New York: Anchor Books, 1953.

Acknowledgments

Writing this book took me far from the usual territory of a geologist. It began as a straightforward story about fossils and strata, but one question led to another, and before long I was delving into unfamiliar subjects such as seventeenth-century theology and philosophy. It was easy to sympathize with Steno, who wrote in his introduction to *De solido,* "I saw that I was wandering in the kind of labyrinth where the nearer one comes to the exit, the greater the circles in which one walks."

Obviously, a person wandering in a labyrinth needs guidance, and I was lucky to be pointed in the right directions by several people whose knowledge was exceeded only by their generosity. Thanks first to Kerry Magruder, a historian of science at the University of Oklahoma and librarian of the university's wonderful history of science collection. At Kerry's invitation, I spent a week in Norman, where I not only picked his brain and perused the library, but also enjoyed discussions with graduate students and other faculty members in the History of Science department.

I am delighted to extend my warm thanks to the Danish Steno scholars who encouraged my work and during my visit to Copenhagen made me feel very welcome. Jens Morten Hansen took time from his duties as the director of the Danish Research Agency to share with me his ideas on the significance of Steno's science and philosophy, and even gave me a brief tour of the local geology. Troels

Kardel, M.D., enlightened me on Steno's genius as an anatomist, to which I have been unable to do full justice in this book. Sebastian Olden-Jørgensen at Copenhagen University patiently explained to me in person and in numerous follow-up e-mails the theological and personal issues involved in Steno's conversion. Here in North America, John Heng at King's College in Ontario shared with me his understanding of Steno's involvement in the scientific and theological controversies of the day.

Much of my library research was done at the Smithsonian's National Museum of Natural History; I gratefully acknowledge the assistance of its staff. Though most of my friends and colleagues in the Department of Paleobiology have hardly seen my face since I began full-time work on this book two years ago, I thank them all for their interest and encouragement on those occasions when I did emerge from the library stacks.

Many other people provided helpful information, advice, and research materials, including: Roger Ariew, Virginia Polytechnic Institute; Peter Barker, University of Oklahoma; Ennio Brovedani, S.J., Istituto Niels Stensen, Florence; Claudine Cohen, Ecole des Hautes Etudes en Sciences Sociales, Paris; Elisabetta Cioppi, Museo di Storia Naturale, Florence; Helge Clausen, Aarhus University; Paula Findlen, Stanford University; Morten Fink-Jensen, Copenhagen University; Alan Ford, Nottingham University; Charly Garbarsch, Copenhagen University; Michael John Gorman, Stanford University; Michael Hunter, University of London; Lawrence Principe, Johns Hopkins University; Ole Petersen, Copenhagen University; Frank Allan Rasmussen, Copenhagen University; Gary D. Rosenberg, Indiana University-Purdue University Indianapolis; Martin Rudwick, University of Cambridge; Jole Shackelford, University of Minnesota; Thomas

Soderqvist, Copenhagen University; Kenneth Taylor, University of Oklahoma; Davis Young, Calvin College; and Lorenzo Del Zanna, S.J., Copenhagen.

In trying to fold all the complexities of Steno's life and the evolution of geological thinking into a single, readable narrative, I had to compress and simplify many things that a professional historian writing for professional historians would no doubt have discussed more completely and rigorously. I hope my sources and consultants agree that I've gotten my facts right, even where I didn't take their advice or had to leave out much of the good material they gave me. If I have erred in places, the responsibility is mine, of course.

Luz Cutler, Charles Cutler, and Margaret Cutler read early partial drafts of the book. Their comments and responses as readers were more helpful than they may have realized at the time. Kerry Magruder, Tim Sullivan, Beth Sullivan, Margaret Cutler, Charles Cutler, and Roger Cutler read complete drafts. I thank them for the careful scrutiny they gave the manuscript and for their always constructive criticisms. Thanks to Zubin Tejani, Amy Cutler, and Nicholas Cutler for their help in the office, doing photocopying and other necessary tasks.

I want to thank my editor at Dutton, Mitch Hoffman, for his good humor and steadfast support throughout the project. He believed that I knew what I was doing, even when I wasn't so sure, but he often turned out to be right. Many thanks to Mitch's assistant, Stephanie Bowe, and to everyone else at Dutton who helped produce this book.

I owe an incalculable debt to my brilliant agent Jody Rein, who nurtured this book through all stages of its evolution: from a mere idea back in the Precambrian, to a book proposal in the Mesozoic, and, finally, after some geologic upheaval, to a finished book today. It could not have happened without her. My thanks also to Jody's assistants and to Agnes Krup and Jenny Meyer.

My parents Charles and Margaret Cutler and my siblings Tom Cutler, Roger Cutler, and Patricia Silber have been a great support always, but especially during my work on this book. Thanks for all your words of encouragement, and for everything you've done for me and my family.

Finally, I want to thank my wife, Luz, and my children Nick and Amy, who saw their husband and father disappear into the seventeenth-century about two years ago. I'm back. Thanks for your faith in me.

INDEX